Springer Biographies

The books published in the Springer Biographies tell of the life and work of scholars, innovators, and pioneers in all fields of learning and throughout the ages. Prominent scientists and philosophers will feature, but so too will lesser known personalities whose significant contributions deserve greater recognition and whose remarkable life stories will stir and motivate readers. Authored by historians and other academic writers, the volumes describe and analyse the main achievements of their subjects in manner accessible to nonspecialists, interweaving these with salient aspects of the protagonists' personal lives. Autobiographies and memoirs also fall into the scope of the series.

Yuan Wang

My Mathematical Life

Yuan Wang in Conversation

Yuan Wang
Academy of Mathematics and Systems Science
Beijing, China

ISSN 2365-0613 ISSN 2365-0621 (electronic)
Springer Biographies
ISBN 978-981-19-3550-3 ISBN 978-981-19-3551-0 (eBook)
https://doi.org/10.1007/978-981-19-3551-0

Jointly published with Science Press
The print edition is not for sale in China mainland. Customers from China mainland please order the print book from: Science Press.

Mathematics Subject Classification: 01A70, 01A25, 11-XX, 65-XX, 62-XX

© Science Press 2024

This work is subject to copyright. All rights are solely and exclusively licensed by the Publisher, whether the whole or part of the material is concerned, specifically the rights of reprinting, reuse of illustrations, recitation, broadcasting, reproduction on microfilms or in any other physical way, and transmission or information storage and retrieval, electronic adaptation, computer software, or by similar or dissimilar methodology now known or hereafter developed.
The use of general descriptive names, registered names, trademarks, service marks, etc. in this publication does not imply, even in the absence of a specific statement, that such names are exempt from the relevant protective laws and regulations and therefore free for general use.
The publishers, the authors, and the editors are safe to assume that the advice and information in this book are believed to be true and accurate at the date of publication. Neither the publishers nor the authors or the editors give a warranty, expressed or implied, with respect to the material contained herein or for any errors or omissions that may have been made. The publishers remain neutral with regard to jurisdictional claims in published maps and institutional affiliations.

This Springer imprint is published by the registered company Springer Nature Singapore Pte Ltd.
The registered company address is: 152 Beach Road, #21-01/04 Gateway East, Singapore 189721, Singapore

Paper in this product is recyclable.

Preface

As far back as the year 1982, I was invited by *Science*, a Japanese magazine, to write *My Mathematical Life* and later *My Memories,* elaborating on my life experience of studying mathematics, which were afterward revised and included in my essay collection known as *Wang Yuan on Goldbach Conjecture.*

For the backbone of my mathematical research, i.e., sieve method and Goldbach Conjecture, Diophantine analysis, applications of number theory to numerical analysis, mathematical methods in statistics, and the study on history of mathematics, I'm too sickly to go into details and could offer merely a few reminiscent words on the topics.

The writing plan, once shelved, would not have been fulfilled without the help of my old friend Li Wenlin and new teammate Yang Jing. My sincere thanks go to both of them. The three of us had a half-day discussion each week. During the 20-plus discussions, they asked questions, I answered questions, and after the discussions, we got on with our work separately. Moreover, Yang Jing took the tedious task of typing to record.

After consultation, the title was chosen as *My Mathematical Life—Interviews with Wang Yuan,* Edited by Li Wenlin and Yang Jing, Wang Yuan Dictation. In the future, if anyone thinks there's more to write about me, they can do, I myself, however, will have no further involvement in.

Beijing, China   Yuan Wang

Introduction

Wang Yuan the CAS academician has reiterated that, for a mathematician, mathematics is of utmost importance. As you could see from the title, *My Mathematical Life: Interviews with Wang Yuan* is an autobiographical collection of interviews featuring Wang Yuan's life experience of studying mathematics.

Goldbach Conjecture, among others, is definitely a highlight in Wang's career. Wang Yuan proved (2, 3) and published his result in 1957, for the first time a Chinese scholar on the leading edge of research on Goldbach Conjecture, one of the famous Hilbert's Problems listed together with Riemann Hypothesis and Twin Prime Conjecture, which though unsolved, could nonetheless serve as incubators for new mathematical ideas. Thanks to its innovative approaches, Wang Yuan's research findings have been guiding younger generations of domestic mathematicians all along their way up to the high mountain while being quoted by his peers across the world. Research on Goldbach Conjecture is definitely an important event in contemporary mathematics history of China and Wang Yuan with his findings has carved out distinct marks in it. In Chaps. 4 and 5, Wang Yuan reviewed his journey of studying the Goldbach Conjecture, dwelling on research motivations, innovative approaches as well as international evaluation. If you ever want to find out how a significant mathematical finding is unveiled and thus gain some inspiration, this book, faithful as it is, would be the scientific document worth reading.

Following his research on Goldbach Conjecture, Wang Yuan moved on to the applications of number theory. He worked with his teacher Hua Loo-Keng to apply number theory to calculation of multiple integrals and established a numerical integration method known as the Hua-Wang Method. Since the 1980s, Wang Yuan worked with Prof. Fang Kaitai, a younger colleague, to apply number theory to mathematical statistics and established the Uniform Design Method widely used in industry, agriculture, and national defense. Collaborative research of Wang Yuan and his partners remains in bloom and has blazed new ground in the applications of the number theory a field usually considered as useless. In Chaps. 6 and 9, We could have a glimpse on Wang's insights into applied mathematics and how he worked on researches in this field. In particular, we could see how the mathematician had always taken the right direction at the right time and thereby secured results of significance.

Mathematicians are more often believed to be working individually and Wang's story of collaborative researches on the application of number theory may offer us a fine example of fruitful win-win cooperation in mathematical research.

Algebraic number theory was extensively used in Wang's joint research with Hua Loo-Keng in establishing the number-theoretic method for approximate calculation of integrals, which brought him back to the researches on pure number theory for a period of time afterward concluded with his book *Diophantine Equations and Inequalities in Algebraic Number Fields*. His findings in this regard have been celebrated among his foreign peers as "a valuable contribution to literature regarding Hardy-Littlewood Circle Method." This part of Wang Yuan's career is to be found in Chap. 8: *Back to Math*.

Wang Yuan places great emphasis on the history of mathematics. His research in this field started from 1985 and major publications include *The Goldbach Conjecture* and *Hua Loo-Keng: A Biography*, the former came as a systematic review of the available papers on the conjecture, while the latter offered a vivid account of the career and life of Hua Loo-Keng, the most important mathematician of contemporary China, and some big moments involving him in China's contemporary mathematics history since the 1920s. *Hua Loo-Keng: A Biography* has been translated into English and published by Springer as its one and only publication on a Chinese mathematician's life story. Chapter 10 *A Sip of Mathematics History* would explain why and how Wang Yuan conducted both of the above-mentioned researches and his understanding therein, offering Wang Yuan's thoughts on mathematics history.

This book, though focusing on Wang's academic career, enthralls readers with the charm of his personal character.

During the interviews, Wang Yuan repeated that people should have "an apt self-assessment," which exactly helped him stay humble. Wang had a successful debut in 1956 when he attracted substantial coverage (for his joint research findings on arithmetic function with Polish mathematicians) in *China Youth Daily* as a young scientist model in China's "March toward Science" program. Instead of indulging himself in the glory, he buckled down and marched toward a greater program: Goldbach Conjecture.

As a well-known mathematician and CAS academician, Wang Yuan however always maintains a low profile. Apart from his job as a CAS researcher, he rarely accepts appointments from other organizations in spite of many urgent invitations received. Due to CAS appointment and elections, he has taken administrative positions like director of the Institute of Mathematics under the Chinese Academy of Sciences, but insisted he would serve only one term for any of these positions. He is the academician applying for retirement at the youngest age and a CPPCC member who proposed not to be reelected. For Wang Yuan, mathematics is always where his heart and soul belong to.

During the interviews, Wang Yuan handed over his reading notes to us. It's a massive 4900-page collection including notes since his college days, telling the hard work he has dedicated to scientific researches.

To quote Karl Weierstrass, "it is true that a mathematician who is not somewhat of a poet, will never be a perfect mathematician." Evidently, Weierstrass was not a poet and he was simply stressing that a perfect mathematician shall be an artist in some way. Wang Yuan is a Huqin and violin player and enthusiast for calligraphy and paintings since school days. This book contains a copy of an autographed pencil drawing completed during his second year at high school, accounting for his accomplishment in calligraphy in later years. Like he said during the interview, "for me calligraphy is science, instead of skills." He explores and practices calligraphy like how he had dealt with Goldbach Conjecture and has thus cultivated a distinctive writing style. Perhaps we would be in a better position to understand and appreciate his calligraphy works as a calligraphy-mathematics integration.

Wang Yuan is a broad-minded person. We would see from this book that, he would rather "laugh off the bad old days", even for the unfair sufferings endured during the Cultural Revolution.

The first of interviews with Wang Yuan took place at the end of spring, 2018. Wang Yuan, at the age of 88, had a 2-hour dialogue with us each time and still impressed us with his clarity and perception. We had the last interview in early 2019. Professor Wang Yuan passed away on May 14, 2021, at the age of 91. The publication of his autography will be a memorial to the venerable mathematician.

Wenlin Li
Jing Yang

Contents

1	**Childhood**	1
	1.1 Childhood Memory	1
	1.2 Resettlement	5
	1.3 Shilongzhai Castle	6
	1.4 Memories of the Primary School	7
2	**High School**	11
	2.1 Memories of No. 2 High School	11
	2.2 No. 6 High School of Nanjing	13
3	**University**	17
	3.1 Turnaround	17
	3.2 Heart and Soul	18
	3.3 Episode	21
	3.4 Self-Study	21
	3.5 Life Beyond School	23
4	**Connection with Number Theory**	25
	4.1 First Year in the Institute of Mathematics	25
	4.2 Goodbye Functional Analysis	27
	4.3 Hua Loo-Keng's Disciple	27
	4.4 The Seminar on Introduction to Number Theory	29
	4.5 Seminar on Goldbach Conjecture	30
	4.6 The Good Help	32
5	**Sieve**	35
	5.1 "To Have Speed, But Also Acceleration!"	35
	5.2 $(3, 4)$ and $(1, 4)_R$	38
	5.3 $(1, 3)$	40
	5.4 By-Products	42
	5.5 Chinese School of Analytic Number Theory	45

6 Applied Mathematics Explored ... 49
- 6.1 Commitment to Applied Mathematics ... 49
- 6.2 Geometry of the Ore Body ... 50
- 6.3 Linear Programming ... 52
- 6.4 Pseudo-Monte Carlo Method ... 54
- 6.5 The Hua-Wang Method ... 57
- 6.6 Book Review of *Applications of Number Theory to Approximate Analysis* ... 62
- 6.7 Motivation ... 69

7 Lost Memories ... 71
- 7.1 Storm Coming ... 71
- 7.2 Regret and Sadness ... 72
- 7.3 Qianjiang Cadre School ... 73
- 7.4 Back to Normal ... 74

8 Back to Math ... 77
- 8.1 Selecting New Area of Research ... 77
- 8.2 Extension of Schmidt's Results ... 79
- 8.3 Book Review of *the Diophantine Equations and Inequalities in Algebraic Number Fields* ... 81

9 Dabbling in Mathematical Statistics ... 85
- 9.1 Origins ... 85
- 9.2 Book Review of *Number-Theoretic Methods in Statistics* ... 88
- 9.3 Implications ... 90
- 9.4 Gentlemen's Agreement ... 92
- 9.5 Coincidence of the End and the Beginning ... 93

10 A Sip of Mathematics History ... 95
- 10.1 Goldbach Conjecture ... 95
- 10.2 Hua Loo-Keng's Outline ... 96
- 10.3 Hua Loo-Keng: A Biography ... 97
- 10.4 Hua Loo-Keng's Mathematics Career ... 100
- 10.5 Done and Dusted ... 101

11 When the Curtain Falls ... 103
- 11.1 Selected Papers of Wang Yuan ... 103
- 11.2 Academic Evaluation ... 105
- 11.3 Popularizing Science ... 107
- 11.4 Enthusiast for Calligraphy ... 109

Chapter 1
Childhood

1.1 Childhood Memory

My family comes from Zhenjiang, Jiangsu Province, yet I was born in Lanxi County, Zhejiang Province on April 15th, 1930, or March 17th of the Gengwu Year in Lunar Calendar, when my father Wang Maoqin, whose courtesy name known as Mianchu, served as the magistrate of Lanxi County. Originally my name was Wang Yuanlan. Since my grandfather had already passed away at that time, my father had few to resort to on his father's side and was helped and funded generously by my grandma's brothers. Thus, I shared the same middle name Yuan with cousins from my grandma's family the Zhus and took the name Lan from my birthplace Lanxi. When I was to attend primary school, Lan was removed and my name has been Wang Yuan since then.

My great-great-grandfather Wang Jiacheng was a famous silk merchant in Zhenjiang of Jiangsu Province. His Laoqingsheng Silk Store, with the headquarter in Zhenjiang and a branch in Yangzhou, used to be the leading silk brand name in those areas. He was a benevolent donor always ready to help others, well known for funding the Zhenjiang Red Boat (rescue boats in the Yangtze River) at that time. His close friend Ruan Yuan, courtesy name known as Boyuan and art name as Yuntai, used to be an influential political figure in Qing Dynasty during the reins of Qianlong, Jiaqing, and Daoguang (1789–1838), successively serving as Governor of Zhejiang, Governor of Jiangxi, Governor of He'nan, Governor-General of Hunan and Hubei, Governor-General of Guangdong and Guangxi, Governor-General of Yunnan and Guizhou, and finally, the Grand Secretary of Tiren Cabinet and Grand Preceptor when he retired. Besides, he was a prominent scholar on mathematics, astronomy, Chinese geomancy, historiography, epigraphy, collation, etc. He was a prolific writer and his book *Biographies of Astronomers and Mathematicians* was the first of its kind in China. At the start of the Taiping Heavenly Kingdom Movement, the Wangs moved from Kaisha, Zhenjiang to Gongdao Bridge, Yangzhou, near the Ruans' house. The two families were close neighbors and good friends so they decided to forge closer family ties through marriage. My great-grandmother was the youngest daughter of Ruan Yuan's fourth youngest cousin. She was a beloved daughter before marriage and a respected lady among the Wangs, which could be attributed to her family background as well as her pleasant personality, wisdom, grace, and elegance.

Ms. Ruan and my great-grandfather Wang Zengqing had six children, among which my grandfather Wang Zhaochang, the fourth eldest child married to Zhu Yuru from Jingde, Anhui, had four sons and four daughters. My father Wang Maoqin, the fourth eldest child, discontinued his study after graduating from primary school and served as magistrate of Pujiang County and Lanxi County after passing the first magistrate recruitment examination as a self-learner in Zhejiang Province. He retired as Chief Secretary of the Academia Sinica.

As long as I could remember, my family had moved to Hangzhou, living in a house at 9, Hehuachitou Community, Qingbomen Gate. My father served as section chief in the provincial government. Apart from bedrooms and living rooms, we had front yard and back yard in the house. In the front yard, there was a water vat under the eaves used to collect rainwater. In the back yard, we kept several chickens. One for each of us and for me it was a black-feathered hen, which was called Old Black Hen by all. We had to kill it when leaving Hangzhou at the outbreak of the War of Resistance Against Japan. I was heartbroken.

My family was better off than other relatives, so my grandmother and my father's two sisters and one brother lived with us. My mother Wang Renqiu came from Suqian, Jiangsu, and was naive and sweet-tempered. My brother Wang Yuanbai, 1 year younger than me and later renamed Wang Ke at school, was brought up by my mother, while I was brought up by my grandmother, who was quite competent and resourceful. Later, I got to know that my grandfather died at his forties. My father, barely 18 years old then, had to work and make a living, while my grandmother managed housework, struggling to support the family together. My father has one elder brother, one elder sister, two younger sisters, and one younger brother. His elder brother and second youngest sister died of illness at their twenties. With my grandmother, he worked hard to raise the rest of the siblings, and his youngest sister and brother even accepted middle school education.

1.1 Childhood Memory

Besides Wang Ke, I have two younger sisters. Wang Yuankang the older one was born in 1933 and renamed Wang Zhiyou at school, and Wang Yuanfan the younger one was born in 1934 and renamed Wang Zhishi at school. I remembered when Wang Zhishi was born, someone told me, "your mother gave birth to a beauty." I yelled "wow" outside my mom's room to remind her that I was still here.

I am the first grandson in the family and was therefore much favored. My grandmother adored me and often shared with me the nourishing bird's nest my father bought for her. When I was four, I went to Qingbomen Kindergarten with Wang Ke. It's said that I was shy. Often I would sit alone at the corner and suck on my clothing, waiting for my family to take me home. In contrast, Wang Ke was poised and acclaimed for that. My father gave me several *Cock* and *Duck* pencils, which I cherished so much that I didn't throw away until they were cracked a few years after my university graduation.

A portrait of Ruan Yuan

Ms. Ruan, Wang Yuan's great-grandmather

Wang Yuan's mother held Wang Yuan in her arms

1.2 Resettlement

Soon after I went to primary school, the War of Resistance against Japan broke out. My family moved to Shuigetang Town, Lanxi County, and stayed in the Jiang's mansion. Wang Ke and I went to the primary school attached to the Jiang's Ancestral Hall. In front of our residence stood Tiejia Mountain, which was bare and barren with no trees. From time to time, I had the impulse to climb up and take a look, but finally, I didn't. I remember that in the school Wang Ke once accidentally punched on my face. As a result, I had a nosebleed at the moment and some more in later years.

As the raging war spread, we proceeded with our escape toward inland by train. I could still remember at one midnight when we were waiting at the railway platform, a KMT soldier was there whipping a deserter. It was tragic.

We went from Zhejiang to Hunan via Jiangxi and spent some time in Changsha. Since my youngest aunt's husband Feng Wenqi and uncle Wang Maoxun both worked with the Liuzhou Southwest Road Authority, we fled on to Liuzhou and stayed in their home. I could still remember the beautiful scenery and inexpensive fruits there. Sometime later the Southwest Road Authority was to send a freight truck convoy to Chongqing and both families took the chance to get to Chongqing via Guiyang. We sat beside the drivers in different trucks respectively and traveled through mountain vistas and blooming greenery. As we passed by the precipitous Gousi Cliff, all of us got off trucks, walked to the mountain foot, and waited on the level ground. I saw some vehicle debris below the cliff.

My mother gave birth to my youngest brother Wang Yuandan on January 1, 1939, and Wang Guang was his formal name. We lived in Chongqing for a while before it was bombed by the Japanese troops. Again we had to move and went further to rural areas. First, we moved to Gele Mountain and lived in one of the newly built suites on the mountainside. Life there had been pretty lonely—there were few woods and no people around except for a few families.

My father then was the director of the KMT Membership Registration Division, Organization Department, KMT Central Committee. Some divisions under the Organization Department, including the Membership Registration Division, were moved to the rural areas in Yuelaichang Wharf, Jiangbei County. After drifting from one place to another, finally my family settled down in Shilongzhai Castle in 1940.

1.3 Shilongzhai Castle

Yuelaichang Wharf was located along the Jialing River and the stockaded village on the mountaintop across the river was known as Shilongzhai Castle, below which there was Zhaojiawan Creek with a group of tile-roofed houses. Both the houses and the village belonged to Zhu Dianchang. The houses of Zhaojiawan Creek had been rented to the Organization Ministry as offices. Zhu Dianchang's family lived at the back of Shilongzhai Castle and the eight or nine rooms in the front were rented to the leadership of the Registration Division.

Shilongzhai was a castle-style building fenced and fortified with stone walls, its 2 stone doorways leading to 20-odd elaborate tile-roofed houses as well as a small pavilion surrounded by longan trees, tangerine trees, tangor trees, and other fruit trees. A dug well adjacent to the North Gate could supply water as shallow as less than one meter or sometimes two to three meters deep. Moss and shrubs grew round the well, of which there were stone steps leading to the bottom. It was said the well never dried up no matter how dry it got.

My family lived in three south-facing rooms in the front yard, the one in the middle was our living room and the other two were bedrooms. We had a kitchen outside.

It was hot in summer and cold in winter in the house due to windows all around. However, we had a large courtyard with four large osmanthus trees in it. When autumn came, the flowers would blossom and scent the air near and far. Standing by the castle on a sunny day, one could catch sight of Beibei and Chongqing Airport. There was a ferocious dog called Black Foal and we were frightened. One day it died somehow.

We led a hard life back then and could only afford moldy rice with lots of bits and pieces in it. My mother would take the four of us siblings to pick and clean up the rice before cooking. We chose cheap cloths to make clothes. Our parents often did the sewing for us by the tung oil lamp at nights. The good old days!

Life was difficult, so we kept getting sick all the time, like malaria, dysentery, scabies, etc. My sister Zhishi got serious dysentery and turned as thin as a skeleton. Both my parents thought she might not make it that time. But gradually she recovered and lived through the sickness at last. The only one bottle of quinine to treat malaria soon ran out. With few doctors or medical supplies in the rural area, blessedly, we survived the disease every time. Each and every adult stayed safe and all of the kids grew up.

My grandmother and Wang Maoxun my uncle went to live with my uncle-in-law Feng Wenqi's family after arriving in Chongqing. They moved to northwest China and we had not met until the victory over Japan. On the face of it, it seemed my mother and my grandmother were polite to each other. My mother called my grandmother "mom" and my grandmother called her "madame". In fact, my mother was not happy. She had never been the decision-maker of the family since she got married, while my grandmother had the whole family in her hand. At that time, my father had to provide for the three under-aged siblings, Wang Maolin my second youngest aunt, Wang Maoxi my youngest aunt, and Wang Maoxun my uncle. Wang Maoan my eldest aunt, though married, lived on a tight budget and had to ask for support from my father. My mother longed for the modern nuclear family lifestyle and felt dissatisfied. My grandmother was smart enough to be aware. When things changed, i.e., the youngest daughter got married and the youngest son got a job, she chose to live with her daughter and the youngest son unmarried in his lifetime. She stayed with her daughter since then. I think it's what real life would be like, instead of a matter of right and wrong.

1.4 Memories of the Primary School

We had to move frequently due to the war. I could not remember for how many times I had transferred from one school to another. When I turned 10 in 1940, I went to primary school officially. Among the turmoil and chaos of war, my father took time to teach me Chinese and mathematics and somewhat managed to maintain my studies. Wang Ke and I joined Zhaojiawan Primary School as fifth graders. The school, actually established by staff of the Organization Department, recruited both children of the staff and children of local residents and farmers and therefore was run on flexible curricula. We took classes every morning and played out in open in the afternoon. I often went catching frogs and fishing and chopping sugarcane with my classmates. I even ventured to grab the snake tail and shake and then found the snake motionless.

We didn't have too many alumni and, to my surprise, I met some of them decades later on different occasions, respectively. Like Yang Zhihong, whose family fled to Sichuan and set up a small pharmacy in Yuelaichang, went to Shanghai Jiao Tong University and worked with Shanghai Textile Science Research Institute after graduation. Like Yang Dingyan, once a soldier in Chongqing, used to be an actress in Art Troupe of the General Political Department of PLA and later transferred to Beijing Municipal Administration of Cultural Heritage. She looked dynamic and refreshing and was nicknamed giraffe for her long neck. Like Liu Weixian, a lady who could do diversified Beijing Opera makeups. Like Ximen Lusha, daughter of Ximen Zonghua. Born in Moscow and embracing western ways, she went to Tsinghua University and later worked with *People's China* magazine. Each of the three ladies was daughter of my father's colleague or colleague's relative. I also came across Xie Zhengrong, son of a tailor, who settled down in Taiwan now. And Yue Anzhong, who fled from Peking to Sichuan and stayed in Chongqing thereafter. Gao Ziman, son of Gao Yuetian the section chief of the Registration Division, was my classmate too. Gao Yuetian was a lifetime friend of my father. They had been caring for each other all the way through their later years. Gao Ziman went to Zhejiang University too and graduated at the same year with me. We were once in the same team during one CPPCC event.

I remember I was slower in learning new mathematical concepts. We moved to the fraction one year. I just could not figure out how come reduction would come before the addition and subtraction of fractions. That was the evidence I didn't actually understand the fraction. I would not memorize formulas mechanically but rather probe into underlying details and get things straight. That's why when I know it, I know it better! For instance, why 1/2 equals 2/4? I came to understand the point perhaps at least half a year later. Suppose a piece of pancake is cut into two or four, then one piece of the former equals two pieces of the latter. Therefore multiplying the numerator and denominator of a fraction with the same non-zero factor does not change the fraction. In this way, I got to understand the rules of fraction operations.

In the winter of 1940, my father borrowed *The Heart of a Boy*, a book for children. At first, he read the book to me and gradually I could read it myself. The loving story struck a chord with me. We need to love and be loved. That's the first novel I've ever read.

In the autumn of 1941, Wang Ke and I went to fetch water from the well in the castle. Wang Guang and my two sisters went as well. Wang Guang fell into the well by accident. Wang Ke and I had played in some artificial ponds in that summer for several times, yet neither of us could float, not to mention swimming. However, I jumped into the water right away and lifted him out. Fortunately, the water was only waist-deep.

In 2005, I traveled to Chongqing twice for work and for both time I visited Shilongzhai. The old houses had been pulled down and a run-down area surrounded

1.4 Memories of the Primary School

by footing of former walls was all that remained of the old time. A few shabby houses had been built after we left and accommodated some farmers' families. The well however stayed the same. I witnessed low water period and high water once, respectively.

In the autumn of 1941, Wang Ke and I transferred to a more formal school known as Gaofengsi Primary School as sixth graders. The school was located across the Jialing River from my home. I was only 11 years old and lived in the school with Wang Ke. At first, my father used to take us home from school every weekend and send us back on Mondays. Later we came back home by ourselves. I remember our headmaster was Zhou Yun. Once I was sick and delirious from persistent high fever. Luckily Mr. Zhou practiced traditional Chinese medicine and gave prescriptions, which finally helped me recover slowly.

I remember my Chinese Teacher could write modern verses and I tried to learn it. I think modern verse means you write an essay line by line.

My maths teacher was Ms. Zhu. She taught us the four fundamental operations and the Problem "Chicken and Rabbit in the Same Cage" was the most typical. For instance, we count 8 heads and 20 legs among the chickens and rabbits in the cage. How many chickens and how many rabbits do we have? I wouldn't apply the formula my teacher taught, because I didn't understand how they came about. It came gradually into my mind that there would be 16 legs if chickens were all we had, so we definitely had rabbit(s) in the cage. With 1 chicken less, there would be 7 chickens and 1 rabbit, we would have 18 legs. With 1 more chicken less, we would have 6 chickens and 2 rabbits and that would be 20 legs. That was the silly deduction I would use to solve this kind of problem. Then I would get back to the formulas in the book, which were merely formalized deduction. I learned to think independently. I was able to solve some of the arithmetic problems with no formulas to apply to. I would be overwhelmed with pleasure and satisfaction whenever I had any new insights.

I would turn to my father with the arithmetic problems I was not yet able to solve. He could always solve the problem, sometimes a couple of days later. But he could not explain the underlying principles. My father was smart and diligent. At that time I knew my father had never got much formal education and was self-taught because my grandpa died early. I read his diary later and got to know that he had no more than primary school education.

I graduated from the primary school in the summer of 1942. I learned how to swim that summer, that was to float in the water and take a few strokes.

Chapter 2
High School

2.1 Memories of No. 2 High School

Wang Ke and I applied for both the National No. 2 High School located in Hechuan County and the High School Affiliated to National Central University located in Qingmuguan Town. My father was unable to be there, so instead Zhu Baochang his colleague in the Organization Department took us to the examination. We went there by wooden boat one time and walked the other. We got admitted to both. We decided to choose No. 2 High School as it took origin from the time-honored Yangzhou High School.

Its headquarter and senior high school section (for boys) were located within the Puyan Temple, the junior high school section (for boys) was located on the Panlong Mountain and at the foot of the mountain, while the high school section (for girls) at the right side of the mountain. There were 108 steps ascending from the bottom to the mountaintop. The executive offices, dormitory, canteen, and some of the classrooms were situated at the mountain foot, while other classrooms were situated on the mountain. Most of the teachers and students fled to Sichuan from lower reaches of the Yangtze River. We felt especially connected to each other. All of us lived in the dormitory and went back home only in summer and winter vacations.

I went to Panlong Mountain in 2005 and it was packed with houses, obstructing views of the mountain. The alleged historic site of No. 2 High School was kept at the mountaintop, however, the houses all around were built afterward and all that remained from the old school were the four trees in the courtyard. The trees were flourishing.

In spite of the poor living conditions, we were enjoying life. No. 2 High School was like a music school. We made erhu, a two-stringed bowed instrument, out of bamboo tubes sawed and covered with snakeskin we skinned from snakes we caught. Almost everyone in the school could play erhu. Music was all around school. I was one of the best erhu players in my class. I like bright and lively tunes and more than that, I love pieces of Liu Tianhua, *Fine Night*, *In Sickness*, *March Toward Brightness*,

Birds Singing in the Deserted Mountains, etc. Our music teacher, surnamed Zhong, was said to be self-taught and good at playing erhu. I like painting too. But copying is all that I could do. Never have I created a single decent piece. The art teacher, surnamed Deng, was an expert at painting ducks. Privately we called him Duck Deng. Moreover, I like calligraphy. Now it seems those hobbies have been highly beneficial to my self-cultivation.

The experimental class Wang Ke and I went to was a part of the junior high school section of No. 2 High School, which, like other high schools in the past, offered both junior and senior high school education. Some courses, like geometry, algebra, physics, chemistry, etc., would be taught in junior high school and again in senior high school. The difference was the same course in senior high school would review old knowledge and become more challenging. The experimental class, by contrast, made a 6-year overall arrangement and offered every course for once only. For 2 years, No. 2 High School had run experimental classes, which only took top candidates, and our class was the second of its kind. The school took the experimental class seriously and invited some well-known teachers for us. Wang Chusheng our English teacher and Mr. Zhang the maths teacher were kind to us. Mr. Wang taught grammar and text in turn, while Mr. Zhang edited and used his own handouts.

Mathematics and English were my favorite compulsory courses. I love the precise mathematical theories and their strict logical deduction, in particular the assumption-assertion-proof pattern in plane geometry. It takes detailed analysis on contradictions in a problem, increasingly deeper insight and sometimes adding auxiliary lines to solve the problem. For me, the whole process of solving problems upon reflection is motivating. I would be excited and satisfied every time when I made great efforts and finally solved the problem. I felt quite curious and uplifted when I started to learn algebra. How could the Four Fundamental Operations, which I regarded as awfully difficult in primary school, become so simple? All you need to do is to set the equation in unknown(s) and solve the equation, then you'll get the answer. Perfectly clear. With lessons on trigonometry, I was glad to see the series of algebraic relational expressions (or identities) of sides and angles in a triangle and their application, e.g., to measure the height of house and the width of river. After I left No. 2 High School, I learned analytic geometry and again I was excited to see some difficult plane geometry problems could be easily settled algebraically, almost no need to labor your wits, like solving the Four Fundamental Operations algebraically. I've already learned the magic and strength of algebra, a lifetime companion and witness of my growth.

Also I liked English because it was a new language for me. Unlike Chinese characters, an English word could be pronounced as it could be spelt, I felt those were the advantages of English. English beginners often find it hard, but for me that all the more motivated my curiosity and desire to explore.

I had won awards in calligraphy, composition and English recitation competitions when I was in No. 2 High School. I didn't like narrative-based courses, for example, geography, history, nature, etc., which I could understand by reading textbooks instead of spending much time on. My total grade, as a result, was merely above average at that time.

I was fond of reading novels, watching films or students' performance, playing Chinese chess and bridge, swimming, etc., in my spare time. In Hechuan, there was a small river known as the Way of Stone Dragon where we frequently went swimming. I enjoyed hiking too. Our class traveled to Gouyu City one spring. I could still remember the castles on the steep mountains and yellow flowers all around. How beautiful it was! Once we went hiking in an autumn and passed by a tangerine orchard, whose owner allowed us to eat as much as we wanted as long as we don't take away.

I went to Gouyu City again in 2005, but things were different.

It's still fresh in my memory that in a scout camping trip in the wild, we dug a hole in the ground and put a pot in it, we cooked and ate and sneakily attacked each other at night, like it happened not long ago.

In Hechuan, there was a small theater showing old movies, like *Lady of Steel*, and sometimes foreign movies, like *Cabin in The Sky*, as well as some performances, which, now seems mediocre, were once tasteful and fascinating for me. In 1943, the Youth Troupe performed their play *Wild Rose* in Hechuan. Student soldiers of the special training class under the Military Commission and the seniors of No. 2 High School worked together to maintain the order. The two sides came into conflict due to misunderstanding, the student soldiers fired shots and killed Zheng Xuepu, a student from No. 2 High School. Both Wang Ke and I were there but fortunately not affected.

Wang Zhiyou, accompanied by me, went to apply for girls' section under No. 2 High School and got admitted in the summer of 1944. The next summer, accompanied by Wang Ke, Wang Zhishi went to apply for No. 2 High School and got admitted too. We four siblings were then in the same school.

My primary schoolmates Yang Dingyan, Liu Weixian, Xie Zhengrong, and Yue Anzhong went to No. 2 High School successively. I have met or got in touch with them decades later. I've exchanged letters with my classmates from the experimental class Hu Shidi, Wang Jiapei, and Chu Jincai, who lived in Beijing, and Feng Shunlie, who lived in Taipei. I frequently meet with CAS academician Dai Yuanben, once a schoolmate in regular class of the same grade.

2.2 No. 6 High School of Nanjing

In 1945, Japan announced its surrender and the victory over Japan was finally won. We four siblings stayed in school for winter vacation and continued with our studies. Since the Ministry of Education stipulated that the state-funded schools to be dissolved and demobilized, we went back home together at the end of the first semester in 1946.

My father resigned from the Organization Department and went to work in the Academia Sinica (or AS) in 1944 and served as chief of general affairs, secretary-general, etc., successively. Our family moved to the dormitory of AS Chongqing in the summer of 1946 and waited to be resettled to Nanjing. My father at that time had gone to Nanjing and taken office there. Liang Sicheng the architect lived across the corridor and each of us two families occupied one room. Every day I witnessed how this old man, frail as he was, either lay in bed reading or sat there typing. I was very impressed with how industrious and hardworking he had been.

My mother took us five siblings back to Nanjing by transport plane. We lived in a house with a garret in the dormitory of Academia Sinica on Chengxian Street. Wang Ke and I transferred to the High School Affiliated to National Society Education College in Nanjing, which was renamed No. 6 High School of Nanjing. We were day students and lived at home.

The astronomer Zhang Yuzhe, meteorologist Zhao Jiuzhang, historians Fu Sinian and Li Ji, and economist Wu Baosan lived nearby. Though led a life of poverty due to rampant inflation, those scientists stayed committed to the scientific research all the time. They are great men. In 1948, from the AS academician qualification documents my father brought home, I read stories of Hua Loo-keng, Chern Shiing-Shen, Chen Kien-Kwong, Su Bu-Chin, and other mathematicians and looked up to them therefore.

In 1947, there was a dispute between our school and the neighboring No. 5 High School of Nanjing and a riot took place, for which I could not recall exact details. The dean of students planned to expel Wang Ke and me from school for it. Luckily, the headmaster Chen Qipang had learned the truth that we were involved for no fault of our own and managed to ensure fair treatment for us two and calmed the situation by recording major demerits on our disciplinary record.

During the 2 years in the new school, I had enjoyed lots of American and British art movies. *Waterloo Bridge*, the first of those movies, tugged at my heart right away. Though I knew too well that the college entrance examination would determine my destiny, I was so obsessed and unable to stop. Later, I watched *For Whom the Bell Tolls*, *Madame Curie*, *The Great Waltz*, *Hamlet*, *Gaslight*, *The Yearling*, and some British movies adapted from novels by Charles Dickens, which left me impressed with the superb acting and breathtaking stories. This may have distracted me from focusing on school work, but still I'm convinced that unconsciously I had benefited and greatly improved my knowledge of art. Definitely, this has exerted a positive and lifelong impact on me. Mathematics and art actually have much in common. Mathematics should be "beautiful" in the first place to be good and the movies helped me identify and appreciate good mathematics.

I graduated from high school in 1948 at the age of 18. Wang Ke and I applied for six universities and we were admitted by the Department of Physics of Anhui University and Department of Mathematics of Yingshi University, respectively. I came to regret at that point I had not worked harder. Sad and lost as I felt, I was however not disappointed with life. I decided to take the examination again the next year to turn around my destiny.

At that time, the KMT was losing the war while people were suffering from serious inflation and social chaos and student uprisings took place one after another. Nevertheless, my father determined to send us to college. After all, it would be better than self-study at home. As a result, Wang Ke and I decided to go to Yingshi University together. I went to study mathematics and he physics.

2.2 No. 6 High School of Nanjing

My father was always serious. We didn't spend much time together, but he had been our role model all along, silently exerting subtle influence on us. His views and insights had been shaping our upbringings. He told us that our grandfather died early because his opium addiction deprived him of sound health and led to poverty in the family, which kept us siblings away from tobacco. Most often my father warned us against politics and factionalism and insisted that we should depend on expertise to make a living and find our places in the world. He stressed repeatedly, "better personal strength than vast wealth." He expected his children would engage in basic sciences, i.e., mathematics, physics, and chemistry. He had got fed up with politics and this is ingrained in my mind. My father was not interested in fame or wealth and never got involved in scramble for vanity. Our family led a simple life. For many times he urged Wang Ke and me to pick mathematics and physics as our majors and this had influenced my life.

When my father saw us off to Yingshi University at the Xiaguan Railway Station, he stared at the departing train for a long time. He was heavy-hearted, completely different from the way he sent us to Gaofengsi Primary School and No. 2 High School. We were lost at sea and had no idea how the wheel of fate would turn.

I've never met with my parents together since then and it was not until 1980, 32 years later, that I met with my younger sister Wang Zhiyou and younger brother Wang Guang in the USA. We made a phone call to our parents in Taipei from Wang Guang's home and we got too emotional to say a word. I met with father after 37 years of departure in the USA. Then in 1995, I met with my youngest sister Wang Zhishi in Taiwan since I headed for university 47 years ago.

I've met with Wang Naiji, my schoolmate from No. 6 High School who settled in the USA. Still, I have contact with Zhao Zhiwei and Xu Xiaoqi, who are a couple. As the deputy mayor of Xichang before retirement, Zhao gave me information of many schoolmates, including Yan Bojin.

Apart from the 5-year boarding in primary and high schools, I've been with my family for only 13 years.

Chapter 3
University

3.1 Turnaround

In autumn of 1948, I attended the Mathematics Department and Wang Ke the Physics Department of Yingshi University in Jinhua, Zhejiang Province. Neither decent school buildings nor adequate equipment and books were offered, not to mention a complete range of courses. I was disappointed. For mathematics majors there was only one course to take known as calculus, taught by Mr. Lai who commuted between Shanghai and Jinhua to give four lessons every week. The textbook he took, The *Calculus* by Sah Pen-Tung, stressed nothing but calculation and was almost like the unchallenging high school mathematics. Chemistry was the only thing that somewhat sparked my interest. I took time to review the high school textbooks, which I felt easy and had deeper understandings. I was more than willing to learn more, yet I had few access to lectures or reference books.

Away from Nanjing, we were in Jinhua, a city so peaceful and quiet like it were countryside. There was no recreation or pastime except for study and reviewing. We could go swimming in the river when it was hot. There was a zongzi (traditional Chinese rice-pudding) store at the gate of our dormitory, selling ham- and red bean-flavored zongzi. We went there quite often.

At that time, my mother's parents had moved from Nanjing to Jinhua and lived at the home of the eldest brother-in-law of their youngest daughter, who was married to Ni Tingsheng. We had visited them.

It was winter vacation two months after we arrived at school. The Academia Sinica was then busy dealing with changes—the general office was relocated from Nanjing to Shanghai. My family moved to Shanghai too. My youngest aunt and her husband Feng Wenqi lived in Hangzhou with my grandmother and my uncle Wang Maoxun. When the winter vacation began, I went alone to Hangzhou to visit them and planned to go back to Shanghai a few days later. Wang Ke stayed in Jinhua.

My father, though disorientated like most people, had one thing for certain. He insisted that Wang Ke and I stay at school and with school and wrote to me that instead of going back to Shanghai I should get back to Jinhua and stay with Wang Ke. I returned to Jinhua right away.

Soon the AS General Office was relocated to Guangzhou and the whole family moved there with my father, who suddenly changed his mind and wrote to us persistently, urging us to leave for Guangzhou and transfer to Sun Yat-sen University. My uncle wrote to persuade us too. Attached to my father's letter were a letter from a leader with the Ministry of Education to president of the Yingshi University asking the latter to take us to Guangzhou and my father's personal note to Qian Linzhao, Acting Director-general and his senior in AS, asking for help for us if we chose to stay.

Apparently the KMT, corrupt and fragile and subject to a crushing defeat as seen, was approaching it doom. Would there be any difference to stay or to leave? And furthermore, Tsinghua University and Peking University were said to have reopened and restarted admission, I thought it better to stay at school till the liberation and reunite with my family, before we took the college entrance examination again and improve the plight. I believed it was right to follow my father's initial idea and made up my mind to stay. Wang Ke listened to me and followed. We stayed in Jinhua together.

One morning in May, we saw the KMT soldiers marched forward in a row on the slope and shot outside of the city. The People's Liberation Army (PLA) marched into town a few hours later. Though poorly dressed, the PLA was highly disciplined and never infringed on people's interest. I admired them at first sight and felt they were vastly different from the KMT soldiers. In those days mathematics was commonly believed to be a dead-end option. I planned to apply for chemical engineering major the next time. Wang Ke and I went to Shanghai in summer to take the entrance examination. We took time to visit Qian Linzhao at the AS. He was packing up books and told us he was to teach in Peking University. He was concerned about us and offered to fund us.

The newspaper reported that both of us had been admitted but did not clarify which university it was. Meanwhile, the Zhejiang Provincial Government announced to incorporate the students of the science and engineering schools of the Yingshi University into Zhejiang University, the top university in south China and the dream university known as Oriental Cambridge University among youth. It just fell into our laps! Of course, we went to Zhejiang University with the school. Zhou Xianyi and Jiang Zhengrong of the Mathematics Department of Yingshi University went there too. Jiang transferred to civil engineering major of Zhejiang University later.

3.2 Heart and Soul

Located in Hangzhou, a beautiful place blessed with a galaxy of talents and natural endowment, Zhejiang University took in a number of famous scholars, like the mathematical analyst Chen Kien-Kwong and geometer Su Bu-Chin with the Department of Mathematics, nuclear physicist Wang Ganchang and theoretical physicist Shu

3.2 Heart and Soul

Xingbei with the Department of Physics, biophysicist Bei Shizhang and geneticist Tan Jiazhen with the Department of Biology, organic chemist Wang Baoren with the Department of Chemistry, etc. As for the Department of Mathematics, we also had analyst Lu Qingjun and algebraist Cao Xihua returned from the USA, topologist Zhang Su-Cheng returned from Britain, and veteran members Qian Baocong, Xu Ruiyun, Bai Zhengguo, Ye Yanqian, Guo Bentie, etc., and assistants like Gu Chaohao, Zhang Mingyong, Lin Zhensheng, etc.... Amazing faculty! I was so lucky to be a part of it.

When I first came to the university, I remember a schoolmate of physics took me to the lab of Wang Ganchang, the specialist of cosmic rays, who was engrossed in his experiment and ignored visitors to his lab. I stood in awe of him.

I had taken few classes in Yingshi University and had to decide whether to restart as a freshman or take Grade 2 in Zhejiang University. The Department of Mathematics of Zhejiang University offered regular education program and normal education program. Out of concern, an old schoolmate named Deng Jinchu, who took the normal education program, advised me to start over. But I was too ambitious to listen. Previously, I chose to stay in Jinhua partly because I was keen to make changes by myself, instead of transferring to Sun Yat-sen University by riding on my father's coattails. I thought I had spent too much time watching American movies to work hard enough and ended up in Yingshi University earlier. I did not make it to a famous university, but it didn't mean I was stupid. Now I'm ready to throw myself into study heart and soul and this time I'd like to see how it would turn out. If again I could not make it, I would transfer to the School of Engineering. Both Wang Ke and I decided to start from Grade 2.

There were merely a dozen students of mathematics in the university, four of whom were in our class, i.e., Zhou Xianyi and I transferred from Yingshi University, Sun Hesheng a student of mathematics in Zhejiang University from the very beginning, and Du Tingsheng transferred from School of Engineering. There were two for Grade 3, i.e., Dong Guangchang and Li Zezhi, and two for Grade 4, i.e., Guo Zhurui and Guo Fangbai. Several students took the normal education program, which suspended new enrollment in 1949. Grade 1 boasted the largest number of students, like 7–8 including Huang Jiqing, Sheng Zhou (Female), Jin Zhengdao, Sun Yulin, etc. In 1950, we had some postgraduate students, Gong Sheng, Xia Daoxing, Hu Hesheng, etc.

I took nine courses at Grade 2. Gradually, I got to know the traditions of the Mathematics Department of Zhejiang University as well as how we would learn maths and take examinations. Teachers gave lectures from memory in the class, they did not take notes with them while the students would take notes, for most courses there were no textbooks or handouts, and few exercises. Most of the exam questions were basically derived from theorems taught in class. Therefore, you could get full marks if only you were able to figure out how the theorems had been deducted. I got it and knew I'd get it done if I worked hard enough. I made up my mind to get rid of my hobbies and throw myself into mathematics heart and soul.

I felt *An Introduction to Series* was the most difficult course, taught by Lu Qingjun based on Chen Kien-Kwong's handouts, which was edited and compiled based on

Konrad Knopp's books and not given to students. Senior students called it the best part of mathematics major in Zhejiang University. It started from convergence of sequences of real numbers and introduction of ε-δ and ε-N. For me, those were brand new concepts which strictly defined infinitesimal and infinite. It was a new world different from the high school maths and calculus, highly perceptual knowledge centered on calculation and skills. I had to change my way of thinking to adapt to the new maths of a different taste and level. I caught a glimpse of its strictness. I remember Zhang Mingyong used to mark almost everything in my exercises incorrect, it took lots of reflection and consideration before I caught on and got used to it. I was told by seniors that mathematics majors in Zhejiang University were in the position to graduate if only they could make sense of uniform convergence. When it indeed came to uniform convergence, however, I felt it was no big deal and readily understood the concept as the selection of δ irrelevant to the points in the interval.

Another difficult course for me was the *Advanced Algebra* taught by Guo Bentie, who would like to go into details and focus on the Theory of Matrices. It was totally different from high school maths. What is a matrix? It looks like a determinant, which represented a number as we learned in high school and just a calculation skills, while a matrix is a rectangular arrangement of numbers following certain operation rules. Likewise, the plane geometry I learned was based on axioms, yet the axioms there were highly perceptual. The axioms concerning matrices involved none perceptual experience, therefore it took a long time to understand.

We also had the *Advanced Calculus*, taught by Lu Qingjun based on William Fogg Osgood's book, with Gu Chaohao as his teaching assistant. It was calculation-oriented and popular among students, including physics majors sitting in on the class. Bai Zhengguo offered *Coordinate Geometry*, based on handouts by Su Bu-Chin. Both of the above were somewhat similar to courses in high school and I found it easy to learn. As for the *Ordinary Differential Equation* course, it was all about calculation skills and felt like the same as calculus.

After many discussions over course reform among the faculty, our department decided to offer the course of *Elementary Number Theory*, in place of *Equation Theory*, taught by Lu Qingjun with self-edited handouts. Since number theory is the study of integers and we took it slow, it was never a problem for me.

With a year of hard work in Grade 2, I was exposed to a new maths world of a different style as opposed to the high school maths. I came to understand the strictness of theorem proving as well as connotation of ε-δ and matrices. I got high marks in each and every course, except for an optional course offered by the Physics Department, known as *Theoretical Mechanics*, a branch of learning to solve problems in mechanics by writing ordinary differential equations in accordance to Newton's Law of Motion and solving them, like solving problems in plane geometry with algebraic equations, or analytical geometry in short. But I often wrote the wrong equations. Considering I was not good at theoretical mechanics and as a result had fear of physics, I found physics was not the right way for me. However, I was interested and confident in mathematics. I was also one of the most noted students in our department. It was not until then I made up my mind that I would be committed to mathematics for a lifetime. Wang Ke's performance in Physics Department was mediocre, so he transferred to Aeronautical Engineering a year later.

3.3 Episode

Chen Kien-Kwong came to teach *Complex Function Theory* for us in Grade 3, in accordance to Edward Charles Titchmarsh's book. Chapter One of the book was about general analysis, so we started from Chapter Two. With my previous training in the series course, I found it not difficult. We were fond of Chen's lectures and listened eagerly to him, who often slipped a few stories. Xu Ruiyun taught *Modern Algebra* based on Bartel Leendert Van der Waerden's book. I came to understand what were groups, rings, and fields. Xu took things slowly, which, in addition to my previous training on Elementary Number Theory and Theory of Matrices, made her lessons easy to understand.

I taught myself in spare time. At the time I studied on my own initiative. I felt self-study was far more efficient and practical than listening to lectures. In fact, listening in class only brought about shallow understanding and it took self-study after class to gain insight into the lectures. More often than not I felt the teachers proceeded too slowly and I had much time in hand for independent study. I learned *An Introduction to the Theory of Numbers* by Godfrey Harold Hardy and E. M. Wright.

It is worth mentioning that during a complex function test, Chen Kien-Kwong asked a question beyond the textbook, i.e., if $z = 0$, what is $e^{\frac{1}{z}}$? The function takes on different limits when z approaches 0 from different directions, I decided that $e^{\frac{1}{z}}$ has an essential singularity at $z = 0$. As the only one in my class who got it right, I was commended by Chen Kien-Kwong. Later, he talked about it for quite a few times in our department and this has won respect for me.

We had a physical exam when the summer vacation was coming soon. According to the X-ray report, I was diagnosed with tuberculosis. Our school arranged quarantine for students with lung disease in a large classroom in summer vacation. I remember a sick student, stressed and distressed, dropped out shortly after that and passed away soon. I didn't care about it and studied hard as usual. A number of exams afterward showed that I was not afflicted with tuberculosis, it was nothing but a misdiagnose.

3.4 Self-Study

Seniors told us that the Mathematics Seminar Program, advocated by Chen Kien-Kwong and Su Bu-Chin, was the highlight of the Department of Mathematics. There were Seminar Type A and Type B. For the former, advisers would assign a mathematics thesis, while a mathematics book for the latter, to each student that he/she would read and lecture on when the adviser would listen and ask questions. There were only four students in my class, so each student had to give 4–5 lectures every semester. It was self-study under tutoring, unlike aimless self-study, which tended to be shaky and unreliable. In contrast to the passive study with the teachers giving lectures and students listening to lectures and taking notes and exercises, the seminar students would study in an active manner and take more initiative at study. At the same time, gap between students widened. It was a transition from a mathematics student to an independent researcher. From the workshop, advisers would identify students ready to overcome challenges and create something new, i.e., students of potentiality to be a mathematician.

I remember Tan Jiazhen, dean of the School of Science, asked me to take more courses when we were selecting courses for the last year. I told him that I'd like to have more time for self-study and he showed understanding and appreciation. At last, I took *Theory of Functions of a Real variable* by Chen Kien-Kwong, *Differential Geometry* by Bai Zhengguo, *Probability Theory* by Lu Qingjun, and *Topology* by Zhang Su-Cheng. Chen Kien-Kwong compiled the handouts given to students of *Theory of Functions of a Real variable* course and got it published later. Su Bu-Chin's book was used in the *Differential Geometry* course. Neither was difficult. Lu Qingjun lectured in English with self-edited handouts on *Probability Theory*, which was modern mathematical theory on the basis of measure theory, quite different from the combinatorial probability learned in high school. Zhang Su-Cheng generally taught with a booklet by Solomon Lefschetz. At that time China fell behind in mathematics and few domestic scientists knew probability and topology well enough. Perhaps Peking University and Tsinghua University were the only two well-qualified enough to offer the two courses. It was for the first time Zhejiang University had the two courses and a real bonus for me as a college student. After 2 years of hard work, it was not too challenging for me.

Lu Qingjun assigned to me an article of over 100 pages by Norbert Wiener on Fourier analysis and A.E. Ingham's masterpiece *The Distribution of Prime Numbers*. I could not understand Wiener's article and therefore put it aside for the moment to focus on Ingham's book. Though misdiagnosed with tuberculosis at that time, I took time to finish Ingham's book and detailed notes during the summer vacation. In my eyes, analytic number theory was breathtakingly beautiful. Of course, Lu Qingjun spoke highly of my book report. Under the direction of Zhang Su-Cheng, I read and reported on part of the thesis series by Samuel Eilenberg and Saunders Mac Lane on algebraic topology and received favorable comments from Zhang. All I knew about this part was nothing more than surface, I didn't really understand the central issue. However, I was quite interested in topology course and felt the Urysohn's theorem on point set topology was amazingly beautiful.

After a year with the seminar, I had built up my confidence on my capability of independent study into mathematics. Some teachers in Zhejiang University at that time used to copy out the thesis they planned to study by hand. Like them, I copied out some articles on topology before graduation, so I could still learn independently for a while in case I was sent to a place with no books or magazines. By that time, I had made up my mind that mathematics would be my lifelong cause.

I suggested teachers give academic report to us before graduation so that we could study independently afterward. I remember Su Bu-Chin was invited to speak on differential geometry and he spoke from Lobachevski Geometry to Finsler Geometry. Zhang Su-Cheng gave a report on homotopy groups of spheres.

I had been impressed by the diligence of teachers in our department, in particular by Chen Kien-Kwong and Su Bu-Chin. At their age, they were still engaged in mathematics workshop with youths. They learned Russian from scratch and learned it well enough to translate and publish mathematics textbooks in Russian. In fact, Chen Kien-Kwong, then in his 50s, and Su Bu-Chin, in 40s, regarded as middle-aged mathematicians nowadays, were seen as very old at that time. I spent lots of time staying with Gong Sheng, Xia Daoxing, Hu Hesheng, Dong Guangchang, Guo Zhurui, and other seniors, talking with and learning from them. Perhaps Sun Yulin and I were the only ones among classmates giving priority to self-study. He was one grade below me and transferred to the Department of Mathematics of Zhejiang University from Shanghai Textile College. We studied in quite different styles. He often borrowed a lot of books and magazines and read really fast, while I read slowly and would not move on until I got it. I'd stop right away whenever I could not understand. For example, I gave up on Wiener's article after a couple of pages, because it took reading of a book on the Fourier Analysis in advance which I had none.

We had schoolmates in Grade 2 named Wei Daozheng, Zong Yuexian (female), and those in Grade 1 named Shi Zhongci, Xu Yonghua, etc. I had interacted and shared my views on maths with them that they enjoyed so much that they forgot to go sleep.

3.5 Life Beyond School

During my first year in Zhejiang University, I joined the violin team and went to see movies once in a while. I remember *New Biography of Heroes and Heroines* was the first one. I had planned to see *Gone With the Wind* in high school but got no chance, and lost my interest when it was shown in Hangzhou. I had gradually given up all of my hobbies and devoted my heart and soul to mathematics.

Besides mathematics classes, I took part in haircut training course organized by the student union and made a little money with the skill to pay for stationery and soaps. Our school had a hydrology training class to train urgently needed professionals for China when I was in Grade 4 and I was asked to check maths homework for the trainees, in this way, I had a higher income to pay for clothes and shoes. I was strict with my students and treated their exercises the way like Zhang Mingyong had treated mine.

Unlike me, Wang Ke took an active part in school life and loved playing basketball, dancing and other activities. He was elected the chairman of student union in his last year of university.

During my 3 years in Zhengjiang University, I was exempt from tuition and boarding fee and therefore made it to graduation at last. Wang Ke and I had been funded once, respectively, by Liu Zhengyuan, husband of our eldest aunt, and Jin Zaixing, husband of our fourth eldest aunt. Qian Linzhao with the Academia Sinica had remitted money for three times. Each time of the above could cover food expense for a month. I would of course keep this in mind forever. However, I would like to overcome difficulties on my own and every time wrote back to ask them not to send money anymore.

My family moved to Taiwan with the AS General Office. Fortunately, the post service did a great job and always got our mails delivered. My family was overwhelmed with surprise, pleasure, and excitement when they heard we went to Zhejiang University. Once my father asked his acquaintances to send us money from Hong Kong as well as compasses and set-squares, etc.

During my school days, I had witnessed the Three Antis Movement, Five Antis Movement, and Ideological Remolding Movement, mostly among faculty. I could remember nothing more than what Lu Qingjun, dean of the Department of Mathematics, said in a general meeting in retrospect, "I have never used public envelopes and letter paper for private purposes." He got through it soon. The next target was Su Bu-Chin, then academic dean, who was asked to account for the alleged corruption issue. Everyone was chanting slogans, so was I. That was the first Combat Meeting in my life. The charges against Su were later dropped. A shadow was cast over my mind and the political campaigns seemed to me like an obscurity and a pack of lies. I had never heard of Su's alleged corruption since then until later I came across *Shu Xinbei's Archives* and read from the book that Su had been framed.

In 1952, thanks to my academic performance, I was assigned to the Institute of Mathematics, Chinese Academy of Sciences on the recommendation of Chen Kien-Kwong and Su Bu-Chin. Before I left school, Chen Kien-Kwong told me in earnest, "you're like our married daughter, Hua Loo-Keng is the top mathematician in China, go follow him and work hard."

Chapter 4
Connection with Number Theory

4.1 First Year in the Institute of Mathematics

I graduated in 1952. In response to the country's shortage of professionals, schoolmates one grade below graduated in the same year. Du Tingsheng dropped out long ago. As for the remaining three in our class, Sun Hesheng and I were assigned by the government to the Institute of Mathematics, Chinese Academy of Sciences and Zhou Xianyi stayed at Zhejiang University to teach. Sun Hesheng's GPA during his 4 years in college was above 90, mine was right at 80 and Zhou Xianyi's somewhere between 70 and 80. Sun Hesheng was not enthusiastic about research. Once he aspired to a teaching career in remote regions but since he had been assigned to the CAS he just took it. Wang Ke was assigned to teach at Nanjing University of Aeronautics and Astronautics.

Graduates assigned to Beijing left Hangzhou on the same train in early autumn and all were picked up by their organizations from the railway station. CAS arranged a one-month training for the newly arrived graduates in its office in Wenjin Street. We learned the speeches of Chen Boda, then the first vice president of CAS, which talked about the work of CPC members in the CAS and their relationship with the non-CPC scientists, the relationship between senior and young scientists, remolding of intellectuals, etc. Chen Boda stressed, "CPC members should be held responsible for the performance of CAS." "How would CPC member do a good job? Would they abuse the authority of the Party, indulge in blind arrogance, boss around under the name of CPC and believe that no scientists could get things done without their consent? Wrong attitude!", "CPC members should partner with scientists, learn from them and help them cope with difficulties in work." "Senior scientists should care for the young generation and the young have respect for the seniors. We should, for the most part, recognize their strengths, learn from them and ask for their opinions."

I could not agree more with the lecture, which was firmly rooted in my mind. Basically, CAS had worked as requested during the 5 years before the Anti-Rightist Movement. The 5 best years for scientific researches. To my surprise, Chen Boda turned into such a leftist afterward, during the Cultural Revolution in particular. I got confused in retrospect.

We went to work with our own institutes, respectively, after the National Day. Institute of Mathematics was located in a two-story building in Tsinghua campus by the south gate. There were four newcomers that year. Besides Sun Hesheng and me, we also had Xu Kongshi from Tsinghua University and He Shanyu from Peking University. A room was allocated to Sun Hesheng and me as workplace and dormitory. Hua Loo-Keng wrote a poem to celebrate the new arrivals, which was written down on red paper with brush and pasted on the door by Wang Shouren. Moreover, Wang threw a feast for us.

At that time, we looked up to the Soviet Union and the passion of learning from them ran high. Every researcher in our institute including Wu Xinmou, Wang Shouren, and Tian Fangzeng, except for Hua Loo-keng, did a crash course of Russian at the Mathematics Department of Peking University. I lived in teacher's dormitory by the Weiming Lake on the campus. After a month of intense training, I had basically grasped spelling and grammar and with dictionaries could read mathematics documents in Russian.

Some in the institute thought us newly arrived graduates were weak on fundamentals and should take corresponding training for a year. Consequently, Wu Xinmou taught us *Advance Analysis* according to G. M. Fikhtengolz's analysis books, Zhang Su-Cheng taught *Differential Geometry* with self-edited handouts and Zhuang Fenggan taught *Theoretical Mechanics*. Duan Xuefu gave the algebra course and he asked us to rewrite a booklet titled *Lectures on Linear Algebra* of Izrail Moiseevich Gelfand in matrix language. Lu Qikeng and Wan Zhexian worked as assistants for the analysis and algebra courses, respectively, responsible for marking our homework. The institute had, in fact, underestimated us and offered what was too easy for us. I often pointed out Wu Xinmou's slips in his analysis lessons.

I followed the old way in university and spent most of my time on self-study. To consolidate the Russian learned hastily, I read for twice Isidor Pavlovich Natanson's *Theory of Functions of a Real Variable*, a book in Russian, and basically finished exercises in the book. I found it clear and accessible. Then I read the *Elements of Functional Analysis* by L. A. Ljusternik and S. L. Sobolev, also a book in Russian. That was the first time I got to know functional analysis and function space. I recalled analytic number theory was my first academic interest in university. Later I learned topology and found it more interesting than analytic number theory, which made too much of skills. But then I found functional analysis, an integration of analysis and algebra, even more interesting and easy to make sense of than topology. Consequently, I was convinced that functional analysis was my best choice.

4.2 Goodbye Functional Analysis

The first year passed soon. By the end of 1953, the institute decided to confirm the research orientation of the four of us, graduates newly assigned to the institute in 1953, and interns who had already been working with the institute. In the old days, we chose our research orientations according to our understanding of maths and personal passion, instead of considering reputation of the mentor or whether the chosen fields were popular or not. Among us, Zhang Liqian took mathematical statistics, Wang Guangyin worked with Hua Loo-Keng on analytic number theory at first and transferred to differential equation later at his own request, and Lu Qikeng took theory of functions of several complex variables. And some made turns at this point, He Shanyu took mechanics, Wang Chuanying went to develop computers, Gong Sheng and Hu Hesheng chose to stay in Shanghai and had Chen Kien-Kwong and Su Bu-Chin as their advisers, respectively. The Institute of Mathematics respected and supported all of these choices. Since the institute decided to set up the number theory group and differential equation group and so was in need of more professionals in the two fields. For the latter, there were Ding Xiaqi, Wang Guangyin, Qiu Peizhang, and Sun Hesheng. For the former, there were Xu Kongshi, Wu Fang, Wei Daozheng, and me.

At first, I was reluctant to join the number theory group and planned to take functional analysis after finishing the book of Ljusternik and Sobolev. Before making the decision, I went to talk with Guan Zhaozhi and Tian Fangzeng, told them that I had finished the book of Ljusternik and Sobolev, and expressed my will to study functional analysis under their supervision. Guan Zhaozhi and Tian Fangzeng expressed their appreciation and responded politely, "I'm flattered. We could work together after you read a little bit more." They had a high regard for Hua Loo-Keng and stood ready to support Hua Loo-Keng when they heard Hua wanted me in the number theory group. So I kissed functional analysis goodbye.

I made the decision in part because research assistant Wan Zhexian working on algebra talked with me frequently. We often walked together. He made it clear that I should take number theory and do research under Hua's supervision. I faltered under his influence and came around to his opinion eventually. I read Ingham's book in college, so already I had some understanding of analytic number theory and knew about its beauty.

4.3 Hua Loo-Keng's Disciple

In 1948, I was thrilled to read in newspaper that Hua Loo-Keng's masterpiece *Additive Prime Numbers Theory* got published by the Academy of Sciences of the Soviet Union. At that time my father was eager for me to learn mathematics. I told my parents, "One day I'll have Hua Loo-Keng as my teacher." They smiled and joked, "will he have you as his disciple?" It was not until 33 years later in New York that my sister Wang Zhiyou told the story to Hua Loo-Keng.

In 1948, from the AS academician qualification documents my father brought home, I read stories of Hua Loo-Keng, a prolific candidate with many books to his credit.

When we four arrived at the Mathematics Institute in 1952, Hua Loo-Keng met with us in his office. He surprised me. How could someone so famous yet so young? He was only 42 years old!

Once he came to my dormitory and asked me a question: how to convert plane quadratic curves into standard form and rewrite the expression using matrices? I got stuck on the question. He said, "how could you ever forget this!" The next day I gave him the answer. All I had to do was to convert real symmetric matrices into standard form using orthogonal matrices.

Hua was engaged in writing *Introduction to Number Theory* before our research orientations were confirmed. He posed a question to us: for each positive integer n, let $p(n)$ be the number of partitions, then

$$\lim_{n\to\infty} \frac{\log p(n)}{n^{1/2}} = \pi\sqrt{\frac{2}{3}}, \tag{4.1}$$

where log denotes the natural logarithm.

Hua gave a report first. He provided an estimate using elementary methods: when $n > 1$,

$$2^{[\sqrt{n}]} < p(n) < n^{3[\sqrt{n}]}, \tag{4.2}$$

where $[x]$ indicates the integral part of x, and then he proved the estimators:

$$p(n) < e^{cn^{1/2}}, \quad c = \pi\sqrt{\frac{2}{3}} \tag{4.3}$$

He asked us to prove: for any $\varepsilon > 0$, there exists a positive $A(\varepsilon)$ such that

$$p(n) > \frac{1}{A(\varepsilon)} e^{(c-\varepsilon)n^{1/2}}. \tag{4.4}$$

Of course from (4.3) and (4.4), (4.1) is handy. Hua asked us to hand in the test paper once finished.

After a few days of efforts, I proved (4.4) and handed in the test paper. Later, I heard Zhang Liqian turned in his, too. It was actually an open-book test. Perhaps Hua was satisfied with my performance and therefore wanted me in the number theory group.

Hua enthused, "Wang Yuan, let's do number theory together. That's a deal!" By then I had thought about it for a while and decided to join the number theory group. So I answered briefly, "Yes!".

That was how connection with number theory got started.

4.4 The Seminar on Introduction to Number Theory

Besides us four new graduates, we also had research assistant Yue Minyi in the group, who was transferred from Zhejiang University in preparation for the establishment of the institute. Yue used to work on analysis and he was responsible for helping Hua in advising us. We often went to ask him questions. We also had Yan Shijian, a Beijing Normal University graduate of 1952 and student of Min Sihe, who came to the group for further study. Later, we had one more member Ren Jianhua, a teacher coming from Northwest University for further study.

Since the winter of 1953, Hua Loo-Keng had personally run two seminars, one on Introduction to Number Theory and the other on Goldbach Conjecture, for half a day once a week, respectively.

Hua was the keynote speaker of the former. He was engaged in writing *Introduction to Number Theory*. Based on his previous manuscript, he had already finished the first six chapters, i.e., The Factorization of Integers, Congruences, Quadratic Residues, Properties of Polynomials, The Distribution of Prime Numbers and Arithmetic Functions, and gave the handouts to listeners before the seminar was set up. I had stayed with Lu Qingjun's Elementary Number Theory course for a year at Grade 2 in Zhejiang University. Lu took it slow and taught merely a fraction of the first six chapters of the *Introduction to Number Theory*, but still it was not easy for me. While Hua finished the first six chapters in a month or so and I found it quite clear and accessible. Evidently, I had grown up so much unconsciously.

Compared with the first 6, the following 14 chapters were more challenging, each of which fell into different categories and some was independent introduction to a certain field of number theory. We were moving much more slowly. From Chap. 7 on, all of the listeners dropped out of the seminar except for the five members of number theory group and the two coming for further education.

From Chap. 7 onward, Hua changed his way of teaching. For each chapter Hua would first write a rough draft, accounting for around 2/3 of the planned length, and lectured accordingly in the seminar before giving to one of us four (Xu Kongshi, Wu Fang, Wei Daozheng, and Wang Yuan) for supplement, so as to complete the chapter as a whole, which then would be handed to Yue Minyi for revision and at last to Hua Loo-Keng to finalize. Yan Shijian and Ren Jianhua did part of the job, too.

I was learning and studying on the sieve method and therefore responsible for Chapter IX The Prime Number Theorem and Chapter XIX Schnirelmann Density. I made the following changes to the draft: in elementary proof of the prime number theorem in Chapter IX, I replaced the Erdös method, which deduced the prime number theorem from Atle Selberg's Asymptotic Formula, with Selberg's method, and Selberg's Asymptotic Formula was proved substitutively by T. Tatuzawa and K. Iseki's approach. Also I included in the chapter S. Shapiro's elementary proof of the Dirichlet's Theorem, which states that there are infinitely many prime numbers contained in the arithmetic sequence. I made quite a few additions to Chapter XIX, too. There were few revisions to other parts I was responsible for.

In this way, Hua trained his students by involving them in writing a book. It's great, far greater than making students read completed books. In this way, we got more dedicated to the whole thing because we were a part of it. It was not like "hands-off herding", or pushing students to write a book aimlessly, which may well result in inconsistency in contents and styles.

At that time, everyone was immersed in working for the socialism cause that no one would worry about who was to take credit for a certain chapter or whether it should be clarified in the acknowledgment. All of us went to great lengths just for the sake of a better book. Hua's original draft was enriched and expanded with some new achievements as well as the supplement and revision by Yue Minyi and us.

The *Introduction to Number Theory* was completed in 1956, and, thanks to the hard work of Science Press, published in 1957. It was a book of around 600,000 words.

Through the training, we got a whole picture of number theory.

4.5 Seminar on Goldbach Conjecture

Hua Loo-Keng and Yue Minyi planned for the Goldbach Conjecture seminar together and carried it out in four parts, i.e.,

1. Schnirelmann density, L.N. Mann's theorem and Selberg's Λ^2-upper bound method;
2. V. Brun's sieve method and A.A. Buhstab's method;
3. Y. Linnik's large sieve and A. Rényi theorem;
4. The method of exponential sum estimate of prime variables, Carl Ludwig Siegel's theorem, and Ivan Matveevich Vinogradov's Three Primes Theorem.

Hua had talked about why he took Goldbach Conjecture as the subject of the seminar. He said, "I'm not expecting you to make any spectacular achievements. I take the Goldbach Conjecture because it's related to each and every important method in analytic number theory. You'll have a good grasp of those methods by learning the conjecture." On the conjecture, he said, "The Goldbach Conjecture is beautiful, we have no method yet to solve it." On the research orientation of number theory group, he stressed, "If you could make sense of analytic number theory, plus a little bit understanding of algebraic number theory, you would be able to extend the results in analytic number theory to algebraic fields." "As for algebraic number theory, learn Chapter XVI of *Introduction to Number Theory*, in addition to two theorems, i.e., Dirichlet unit theorem and Dedekind discriminant theorem, then you'll be ready to work while learning." That's the analytic algebraic number theory, which had never been put in place as my research topic until I carried it on 30 years later. But that's another story.

4.5 Seminar on Goldbach Conjecture

In a speech on the first days of the Institute of Mathematics, Hua emphasized, "If you are to work on the analytic number theory, I urge you not to work with Vinogradov's method (referred to the method of Weyl sum estimate), you should learn Linnik's method and Selberg's method."

Hua told me repeatedly, "To be a real master means going deep on one subject instead of going wide on a crop of subjects." Speaking of his transfer from number theory, he said, "It dawned on me that Vinogradov's method for estimating Weyl sums could not be improved any further, neither could the principal order be improved nor the secondary or logarithmic order be removed. I would not have transferred if any essential improvement would ever be possible. I would have worked on the Weyl sums right along." "I've been witnessing Vinogradov's improvement of his method all the while before we made a conclusion together." The "conclusion" refers to Hua's article published in 1948 and Vinogradov's monograph. Hua's result turned out to be stronger. There was not more progress in this field until British mathematician T. Wooley made another improvement decades later.

Hua told me, "Maths results rarely leave any traces in history. Sometimes the whole subject ended up in obscurity. The projective geometry was widely followed when I was young, but how much of that has been left?".

His words, though engraved in my mind, was beyond my understanding at the time. It was many years before I came to understand his insight.

Hua proposed that the Mathematics Institute acted to edit and publish two monograph series: the series Type A would be a collection of academic works on personal research results, Type B would be devoted to systematic introduction of a certain field, aiming to make the field accessible to young mathematicians. The program was widely supported in the Institute and beyond.

Hua planned to compile the achievements of the Goldbach Conjecture Seminar, when all its four parts were completed, into collected papers and publish in the series Type B or *Advances in Mathematics*, a magazine established in imitation of the Soviet Union. At that time, international writing on number theory covered only some part of above four aspects, therefore the program was appealing.

Part one as planned was based on papers of Shapiro and Warga, which proved "every sufficiently large integer is expressible as a sum of at most 20 primes." In addition, other materials were included, e.g., Mann's theorem, which was discussed according to Chapter II of *Three Pearls of Number Theory* by A. Khinchine. We four graduates joining the institute in 1952 and 1953 gave lectures at the seminar in turn, while Hua kept asking questions to go through and get to the bottom of every detail. The speakers, faced with the fusillade of questions, often stood and thought long and hard by the blackboard, which was dubbed "hanging by the blackboard". The seminar moved slowly, but the attendants learned a great deal.

Since the Math Institute of Academia Sinica moved all of its books and magazines to Taiwan, there used to be no books or magazines in our institute when it was established. Fortunately, Hua Loo-Keng brought from America quite a few books, magazines, and offprints. Moreover, he left some money in the USA as prepaid subscription for several magazines of the American Mathematical Society. In this way, staff of our institute had the freedom to borrow the new magazines sent from the USA. Borrowers should register in a notebook Hua put in his office before taking away the books, magazines, or offprints and crossed out corresponding records when returning the prints.

Hua had written a manuscript titled *Analytic Number Theory*, containing elementary proof of Selberg's Λ^2-upper bound method and prime number theorem. From his manuscript, we were lucky enough to read these latest achievements among other first readers worldwide.

In those years, China had been learning from the Soviet Union in all aspects. Western teachings were mistakenly criticized in some fields. Although Hua knew too well the world-leading mathematics of the Soviet Union, due to his understanding over the maths worldwide as a whole as well as his knowledge and time-tested professional rigor, he had been paying due respect to and learning from math achievements of both the Soviet Union and the Western countries. Consequently, the math community in China, the Mathematics Institute in particular, had taken an appropriate attitude toward learning from the Soviet Union.

The Goldbach Conjecture seminar moved to discussion on Part Four when the first part was finished. We lectured on the *Introduction to Modern Prime Number Theory* by T. Estermann. In 1956, the Hundred Flowers Movement was announced, followed by the Anti-Rightist Movement. The seminar was thus left aside, not to mention publishing the series Type B.

4.6 The Good Help

Estimation of Exponential Sums and Its Applications in Number Theory is a monograph written for the German series Encyclopedia of Mathematical Sciences. At first, Hua thought writing books too time-consuming and had no plan to accept the invitation of writing the monograph. In contrast, Guan Zhaozhi regarded it an honor to do so and talked him into accepting.

Hua had been making preparations for the writing as early as 1953. First Yue Minyi singled out all articles concerned from *Mathematical Reviews of America* and Zentralblatt für Mathematik in West Germany and handed to Guan Gulan to make note cards. Hua got on with the writing in 1955.

Hua invited Yue Minyi and me as his assistants, maybe he thought I was kind of helpful. Like writing *Introduction to Number Theory*, Hua wrote 2/3 of the book, supplemented by Yue Minyi and me and finalized by himself. I had made great efforts to make sense of the sieve method so I was responsible for helping him in this part.

The book, translated into German by two professors, was published in 1959.

I worked hard as Hua's assistant like I were working for myself and I learned a great deal. I would transcribe each chapter with great care once the chapter was finished and handed it over to Hua for review.

Through the book writing, I got to know the circle method, the method of exponential sum estimate and their applications, the history of analytic number theory, and had a further understanding of Hua's writing style. Hua Loo-Keng would avoid a

4.6 The Good Help

laundry list book, instead he would outline the proof of main methods and theorems, that is to say, explain the profound in simple terms. He had a clear picture of the subject and rarely referred to the note cards while writing. Since I wrote my part in English, I had gained the new skill of writing maths papers in English when the book was completed. I didn't expect to speak of my work in Hua's Book. Therefore, I didn't put advice on sieve method into the book. I didn't know Russian mathematicians introduced my work on the sieve method in the appendix until the Russian version was published.

Yue Minyi went to work for the Operations Research Division in 1958 and I became Hua's chief assistant. I would proofread his manuscript and give advice and supplement. I was careful and cautious at work and would not expect any favors in return. In this way, I learned a lot from him. After working with the number theory group for years, I was sure that research on number theory would be my lifetime career.

Chapter 5
Sieve

Interview

In 1956, you published your first article in a Polish journal and your second paper in 1958. Many newspapers, among them the China Youth Daily, devoted a full-page article in 1956 to your collaboration with foreign mathematicians and your publication abroad. Your first paper was recognized by the mathematical community, which is not easy for many young people who are new to scientific research. Can you tell us about the process of completing your first paper?

Goldbach Conjecture is one of the most famous and difficult problems in mathematics, and it is particularly important because it can be used as a model to bring new methods, concepts, and theories to mathematics. How did you choose such a difficult and important problem to start your research and make important progress? Did you and Chen Jingrun collaborate in this area?

5.1 "To Have Speed, But Also Acceleration!"

I studied quickly, and in about 2 years I had read S.Shapiro and J. Warga's article; Hua Loo-Keng's manuscript on analytic number theory; A. Kinchine's *Three Pearls of Number Theory*; T. Estermann's article on the V. Brun's sieve method; Estermann's book, *Introduction to Modern Prime Number Theory*; A. Chudakov's book, *An Introduction to Dirichlet's L-Function Theory*; Y. Linnik and Chudakov's article on analytical proof of the "Three Primes Theorem"; A. Renyi's article on Goldbach Conjecture; Hua Loo-Keng's *Additive Theory of Prime Numbers*, with A.I. Vinogradov's *The Method of Trigonometrical Sums in the Theory of Numbers*. Suffice it to say that I have read all the required literature in the program for the seminar on the Goldbach Conjecture, except for the fact that I could not find A.A. Buhstab's article.

I was very eager to get something out sooner rather than later, and the feeling was urgent. If nothing can be done, I cannot have a place in the mathematical community. On the other hand, Wan Zhexian, Lu Qikeng, Hu Haichang, Gu Chaohao, Gong Sheng, and Xia Daoxing from Zhejiang University had already published articles, which added to my sense of urgency.

At that time, Hua Loo-Keng had the opinion that anything that could be done with the Buranian sieve method could be improved by the Selberg Λ^2-upper bound sieve method. If this is correct, then by estimating the lower bound of the sieve function by the Λ^2-method, wouldn't that improve the result on Goldbach conjecture obtained by Buran's method! But using the Λ^2-method, I could not estimate the lower bound of the sieve function, and the problem took me more than a year, and I was still at a loss. Later, I came to understand that not where Brun's sieve method could be applied, Selberg's sieve method would necessarily work.

In 1955, the Polish mathematician Kazimierz Kuratowski came to Beijing for a visit. He brought Hua Loo-Keng some offprints of the Polish mathematician's papers, among which was an article on number-theoretic functions by W. Sierpiński and A. Schinzel. Their final result was a proof by Schinzel that for any given h nonnegative numbers $a_1, \ldots, a_h$, and $\varepsilon > 0$, there exists n and n', such that

$$\left| \frac{\varphi(n+i)}{\varphi(n+i-1)} - a_i \right| < \varepsilon. \tag{5.1}$$

together with

$$\left| \frac{\sigma(n'+i)}{\sigma(n'+i-1)} - a_i \right| < \varepsilon, \ 1 \leq i \leq h, \tag{5.2}$$

Here, $\varphi(n)$ denotes the Euler function and $\sigma(n)$ denotes the sum of the factors of n.

Hua Loo-Keng discussed their article with me. He asked what would happen if the twin prime number conjecture was right. Of course, there would be an infinite number of primes p make

$$\left| \frac{\varphi(p+2)}{\varphi(p)} - 1 \right| < \varepsilon \text{ with } \left| \frac{\sigma(p+2)}{\sigma(p)} - 1 \right| < \varepsilon.$$

This immediately led me to the idea that the problem could be handled by the Buranian sieve method; a similar twin almost prime conjecture can be proved by simply replacing the primes with almost primes, i.e., integers whose number of dissimilar and identical prime factors does not exceed a given bound, where we can also assume that the prime factors of the almost primes are greater than a certain bound. Thus, it should be certain that the sieve method can handle this problem, wherein by a parallel extension of the two-dimensional sieve method from the literature to the higher dimensional sieve method, i.e., for each prime, sieve away k residue classes, here $k > 2$, we can improve Schinzel's result. That evening, I wrote a draft outline of an improved Schinger's result using the Buranian sieve method. I proved that there

5.1 "To Have Speed, But Also Acceleration!"

exist positive constants $c_0(\vec{a}, \varepsilon)$ and $x_0(\vec{a}, \varepsilon)$, which depend only on the nonnegative vectors $\vec{a} = (a_1, ..., a_h)$ and $\varepsilon > 0$, such that in any interval $1 \leq n \leq x$ the number of integers satisfying (5.1) is greater than

$$\frac{c_0 x}{\log^{h+1} x} \qquad (5.3)$$

whenever $x > x_0$.

I submitted this result as an article to the *Acta Mathematica Sinica* and translated it into English, and asked Kuratowski to bring it back to Sierpinski and Schinzel. Soon we received a letter from Sierpinski with two articles by Schinzel. In the papers, in addition to the proof of (5.3), it was shown that the number of integers in the interval $1 \leq n' \leq x$ that fit (5.2) also has the estimate (5.3). The letter expressed the intention that Schinzel and I should collaborate in the publication of these two articles, respectively, in the Polish journals *Bull. Acad Polan Sci.* and *Annales Polan Math.* Hua Loo-Keng and I agreed. I received back the manuscript that had been submitted to the *Acta Mathematica Sinica*. These two articles, in which I collaborated with Schinzel, were published in Poland in 1956 and 1958, respectively.

The manuscript I submitted to the *Acta Mathematica Sinica* was reviewed by Min Sihe of Peking University. He started the "Speciality in Number Theory" at Peking University, and enrolled four students. Min encouraged them to communicate more with the number theory group of the Institute of Mathematics and to learn more from Hua Loo-Keng. The number theory group of the Institute of Mathematics also regarded Min as their teacher and often asked him for advice. There was a close relationship between them. Pan Chengdong, Yin Wenlin, and Shao Pinzhong, students majoring in number theory at Peking University, often came to the Institute of Mathematics to attend the seminar on Goldbach's Conjecture, and were nurtured by Hua Loo-Keng. Another student, Hou Tianxiang, came to the Institute of Mathematics less often.

Shao Pinzheng also independently improved on my manuscript and obtained the same result as Schinzel, i.e., extending my results on $\varphi(n)$ to $\sigma(n)$. In reply to Sierpinski's letter, I also informed them of Shao's results. Thereupon, Shao also published an abstract in *Bull. Acad. Polon. Sci.* in Poland.

By then I had read Renyi's article and knew that the proposition was provable when the twin prime pairs in the twin prime conjecture were substituted with pairs of a prime and an almost prime. Renyi used the linear sieve method. By extending his method in parallel to the higher dimensional sieve method, we can change the integers n in Schinzel's and my results to prime numbers p. In other words, I prove that, given the nonnegative vectors $\vec{a} = (a_1, ..., a_h)$ and $\varepsilon > 0$, there exist positive constants $c_1(\vec{a}, \varepsilon)$ and $x_1(\vec{a}, \varepsilon)$, such that in any interval $1 < p \leq x$ the number of primes p satisfying

$$\left| \frac{\varphi(p+i)}{\varphi(p+i-1)} - a_i \right| < \varepsilon, \ 1 \leq i \leq h \qquad (5.4)$$

will be greater than

$$\frac{c_1 x}{\log^{h+2} x \log \log x}. \tag{5.5}$$

whenever $x > x_1$.

Similarly, by replacing $\varphi(n)$ in (5.4) with $\sigma(n)$, the result still holds. I wrote up this result in a paper that was published in the *Acta Mathematica Sinica*, 1958, 1.

When I wrote my first article, Hua Loo-Keng was delighted. He had said, "Not many professors in China have written articles of this caliber." But when I made the second article, he was not happy. He said:

To have speed, and also acceleration!

By "speed", it means producing results. By "acceleration", it means that the quality of the results should be constantly improved. He said to me seriously, "You have to attack big problems."

At that time, the Central Government was calling for a "march to science". China was very backward in science and technology. It was a big event for our own young mathematician to collaborate with foreign mathematicians and publish two articles abroad. Many newspapers reported on this event, especially the *China Youth Daily*, which devoted a full page to it. There was also a picture of a letter from Sierpinski, which caused a sensation. In fact, I was treated as an accomplished example on the march to science. This incident made me accidentally famous throughout the country.

I was cool with the newspaper publicity and didn't take it to heart. I was fully aware to myself that it was only a minor success by chance. But the words of Hua Loo-Keng caused me to think deeply.

5.2 (3, 4) and (1, 4)$_R$

Hua Loo-Keng probably wanted me to use the sieve method to try an improvement of existing results on Goldbach's conjecture. I don't know how I can apply the Λ^2-sieve method to this problem. I was anxious to see the only unread material from Goldbach's Conjecture seminar program—Buhstab's article. I could not find an early Russian journal in this country. With great difficulty, I managed to find an introductory article by R.D.James in an American magazine, which mentions Buhstab's work, but without details.

At this moment, I heard that the CAS library had imported a collection of old Russian magazines. I immediately went to the CAS library in Wangfujing to borrow them. The magazines had just arrived and had not yet been catalogued, and were piled up on the floor of the stacks. Fortunately, the caretaker was sympathetic and agreed that I could look at them on the floor in the stacks. I spent a whole day copying down two articles by Buhstab.

5.2 (3, 4) and (1, 4)$_R$

I quickly learned that Buhstab's method is an identity, which can also be seen as a recursive formula for the sieve function. With each recursion, the upper and lower bound estimates of the sieve function can be improved somewhat.

The first was Brun's dramatic improvement of the Eratosthenes sieve method in 1919, which led to the demonstration that every sufficiently large even number is the sum of two integers whose number of prime factors does not exceed 9.

We will write this result as (9, 9). Similarly, we can define (a, b). Brun's results and methods have been improved by several mathematicians:

(7, 7) (H. Rademachel, 1924)

(6, 6) (T. Estermann, 1932)

(5, 7), (4, 9), (3, 15), (2, 366) (G. Ricci, 1937)

Buhstab first used Brun's method to work out the upper and lower bounds of the sieve function, which we call the initial value of the sieve function, and then used his recursive formula to improve it step by step, thus improving the results obtained by Brun's method alone. He proved that

(5, 5) (Buhstab, 1938)

(4, 4) (Buhstab, 1940)

The problem is already obvious, as long as we can calculate the upper bound of the sieve function by the Selberg method and estimate the lower bound of the sieve function by the Brunian method, using them as initial values, so that we can improve (4, 4) by recursion using the Buhstab's method. The only ready-made result is the "maximum" value of the sieve function estimated by the Λ^2-method, which cannot be used as an initial value yet.

By then, I was familiar with the seive technique. In layman's terms, the technique was already on the tips of my fingernails and would blossom with the slightest stimulus. I remember Hua Loo-Keng casually saying, "Why not use the Selberg method to calculate the upper bound on the sieve function in each range?" Soon I used the Λ^2-method to arrive at an upper bound estimate for the sieve function at each range, added the lower bound estimate for the sieve function from Buhstab's article as the initial value of the sieve function, and iterated Buhstab's method recursively a few times to get a better lower bound estimate for the sieve function than (4, 4), but not quite up to (3, 3). This is really tricky! Fortunately, the two-dimensional sieve method was used here, so I proved that

$$(3, 4) \tag{5.6}$$

Using this method, under the Generalized Riemann Hypothesis (noted as GRH), I prove (1, 4). We write this conditional result as

$$(1, 4)_R \tag{5.7}$$

This improves on the results (1, 6)$_R$ published by Esterman in 1932.

I wrote these two results as two articles and submitted them to the *Acta Mathematica Sinica*, and they were published in issues 3 and 4 of 1956.

The first article earned me a premium fee of over four hundred and eighty dollars, and the second article was paid over three hundred dollars. At that time a college graduate's monthly salary was only about fifty dollars, and this was certainly a large income. I spent over three hundred dollars to buy an Omega watch and to have a tweed coat and a suit made.

5.3 (1, 3)

Hua Loo-Keng heard a miscommunication somewhere that a mathematician in the Soviet Union had proved (1, 3). He asked me how I got it right. The result was so far off! I hurriedly asked Sun Hesheng, a classmate from Zhejiang University who was studying in the Soviet Union, to help me inquire about it. Sun Hesheng and Sheng wrote back with a handwritten copy of A. Vinogradov's article on (3, 3) and (1, 4)$_R$. I quickly figured out that he had obtained the lower bound estimate for the sieve function directly from the Λ^2-method, plus an analytical estimate. The strength of this method should be the same as the strength of my method for proving (3, 4). After careful verification, I found a gap in his proof of (3, 3). I wrote to Vinogradov and told him about it. A year later, after adding new ideas, he proved (3, 3).

In the number theory group, they know each other who is doing what and reading what articles. Yue Minyi asked me to pay attention to a couple of articles on the sieve method by P. Kuhn. I looked at the results of Kuhn's article in the *American Mathematical Review* and learned that he proved that there exist infinitely many x, such that $x^2 + 1$ can be represented as the product of no more than 3 prime numbers. But doing this with the proof of (3, 4) requires replacing 3 in Kuhn's result with 4, so there must be a new idea in Kuhn's article. But unfortunately, I can't find any article by Kuhn in China.

Hua Loo-Keng wrote to P.Turan for this purpose and asked him to help to get ask for an offprint of Kuhn's paper. Photocopying technology was not available at that time, and Turan Tulang sent us a photocopy of Kuhn's paper. I soon learned that Kuhn's idea was to apply a combination of the upper and lower bounds of the sieve function to obtain stronger results. Regarding Goldbach's conjecture, he proved (a, b), where a + b ≤ 6. I combined the method of proving (3, 4) with Kuhn's method, which I had improved, and immediately proved

$$(3, 3), (a, b)(a + b \leq 5), (1, 3)_R. \tag{5.8}$$

In fact, it was very close to proving (2, 3), but a further refinement of the sieve function is needed, i.e., iteration with the Buhstab method. This requires numerically computing some simple integrals. The Mathematical Institute happened to have a

5.3 (1, 3)

mechanical desktop computer, with which to do numerical calculations, I finally proved

$$(2, 3) \tag{5.9}$$

I wrote up these results in two summary articles that were published in the Science Record in 1957.

"I can't believe you made a result in Goldbach's Conjecture itself, " said Hua Loo-Keng to me with great pleasure. "It would be nice if you could go further. If you don't go up, that's probably your life." True to his word, I couldn't imagine that I stopped at 26 and indeed stopped making progress in tackling difficult problems after that.

Inspired by the work of Pan Cheng-dong, in 1962 I replaced GRH in the proof of $(1, 3)_R$ with a "mean-value formula" for the prime numbers in an arithmetic series. This formula was independently proved by E.Bombieri and A. Vinogradov in 1965, thus completing the proof of (1, 3). In 1966, Chen Jingrun published a summary article of (1, 2), and in 1973, when the full proof was given, Scientia Sinica asked me to review the manuscript. I wrote down that "no errors were found in the proof, " and Chen's article was published.

Problems of great importance in number theory are rare. It takes strength to make progress on these problems, but opportunity is also very important. It is something that can only come along once in a lifetime. I should say that I was very lucky; I was able to work next to Hua Loo-Keng, who happened to take the initiative to study Goldbach's conjecture. After I had proved (2, 3), he had been pleased with his initiative to study Goldbach's conjecture. He had said, "What would have happened if I had posed the Fermat problem?" At that time the Fermat problem was very quiet, and it was only after the 1970s that it suddenly came alive again.

I had already received general attention in the national mathematical community. I have seen the textbook "Theory of Numbers" by Buhstab, published in 1960. (2, 3) had been listed there as a theorem. I was also one of the young mathematicians most frequently publicized in the newspapers; on January 20, 1956, Hua Loo-Keng published an article in the China Youth Daily, *To the Youths Marching to the Fortress of Science*, in which he praised, without naming names, some young mathematicians from the Institute of Mathematics, among whom the most prominent one was based on me. He wrote: "There are mathematical papers that break through difficult barriers, papers that greatly exceed the level of the "foreign doctors" before liberation, " and "as we all know, once the difficult barriers are broken, the rewards come rolling in."

A few decades later, I read my father's *Autobiography* and learned that in the 1960s, my high school classmate Wang Naiji had seen my article in the United States and asked a professor at Princeton University about it. The professor replied that this is a promising young mathematician. Wang Naiji wrote to my father and told him about that. This was the first time my parents learned that I was a student of Hua Loo-Keng and my current status. Later, at the CTRC conference, my father had talked about me when he chatted with Chern Shiing-Shen. By then, Chern Shiing-Shen already knew my name and gave the same comment.

My own feelings are mixed. On the one hand, I already have some influence in the Chinese number theory community, and even in the mathematics community. On the other hand, how long could this advantage be maintained? After all, all my work was just an application of the sieve method, working on a very narrow surface. I had no method of my own, and the possibility of improving Goldbach's conjecture further down the line had been slim. My mind was still unsettled, even rather empty.

I realized that my work was "straightforward" and did not involve more mathematical areas. I had worked a bit more on topology and functional analysis than undergraduates, but it was not the same thing as mastering these studies and being able to use them as a flexible tool for myself. I realized that my first priority was to expand my knowledge and pioneer work in other areas.

In 1957, the "Anti-right Movement" was in full swing, I was sent down to Nijiazhuang Township, Gaocheng County, Hebei Province, for "labor training".

5.4 By-Products

In 1958, I returned to the Institute of Mathematics from the countryside. By this time there was a boisterous atmosphere. There were many young advanced teachers coming to the Institute.

The Goldbach conjecture seminar has long since ended, but the number theory group still has one. I learned that they had just finished reporting on D.A.Burgess's 1957 paper on least quadratic positive non-remains n_p modulo p, in which he proved that

$$n_p = O\left(p^{\frac{1}{4\sqrt{e}}+\varepsilon}\right), \; \varepsilon > 0,$$

Here the constants associated with "O" depend only on ε. This is a better result than Vinogradov's

$$n_p = O\left(p^{\frac{1}{2\sqrt{e}}+\varepsilon}\right).$$

There has been a great improvement. I immediately read the article. I knew that he had used A. Weil's result on analogue of Riemann conjecture for L-functions over finite fields to improve the estimation of the corresponding character sum of quadratic characters modulo p. I realized that if I could improve the corresponding character sum of all characters modulo p, then the estimation of the least primitive root $g(p)$ modulo p could be improved. I immediately started an intense research.

Regarding the problem of least primitive roots, I have only come across the results of Hua Loo-Keng written on Chap. 7 of *Introduction to Number Theory*. So, on the one hand, I estimated the corresponding character sum of complex characters, and on the other hand, I sought out and read all the articles on least primitive roots. In particular, I noticed the article by P. Erdös and Shapiro, who used the sieve method

5.4 By-Products

to inprove Hua Loo-Keng's results

$$g(p)=O(2^m \sqrt{p})$$

into

$$g(p)=O(m^c \sqrt{p}),$$

Here c is an absolute constant and $m = \omega(p-1)$ the number of the dissimilar prime factors of $p-1$. The worst estimate of m is $m=O(\log p/\log\log p)$. I also note that N.C. Ankeny's result under GRH

$$g(p)=O(2^m \log^2 p \, \log^2 (2^m \log^2 p)),$$

Of these, Ankeny did not use the Brunian sieve method.

The Brunian sieve method used by Erdös and Shapiro is unlike any sieve method I have ever worked with before. Instead of sieving some residue classes of all the prime numbers in an interval, it sieves only the residue classes of the prime numbers in a small set of prime numbers. Since the number of elements in this set of prime numbers is "very small", we call this sieve method "sparse sieve method". The Λ^2-method does not work.

Regarding Burgess' estimation of character sum, I was still unable to generalize to complex estimation. So, I went to ask Hua Loo-Keng about it. He said, "It seems that H. Davenport has an article on characteristic sums that might be relevant, check it out." I immediately consulted Davenport's article published in 1939, and I quickly extended Burgess' result to estimate the character sum corresponding to the complex character modulo p and then used the " sparse sieve method" to get the expected result

$$g(p) = O\left(p^{\frac{1}{4}+\varepsilon}\right) \varepsilon > 0, \tag{5.10}$$

Also under GRH, the "weighted sparse sieve method" was applied to improve the results of Ankeny to

$$g(p)=O\left(m^6 \log^2 p\right) \tag{5.11}$$

I told Hua Loo-Keng about these two results. He was quite surprised, and he made a very complimentary remark about me. I submitted an abstract of the paper to the Science Record, which was published in 1959. It was unexpected that the paper would be cited in some information science literature a few decades later. It was a "by-product" kind of job. The whole thing only lasted a month.

In 1958, the Chinese Academy of Sciences established the University of Science and Technology of China (USTC), located in Yuquan Road, Beijing. Hua Loo-Keng was appointed as the head of the Department of Applied Mathematics. He personally gave a basic course in advanced mathematics to the first-year students, which I taught together with him.

By this time, though, my research field had moved away from the sieve method, I still paid attention to its progress. In 1963, I switched to teaching the students of the "Number Theory and Algebraic Specialization, " where I was responsible for teaching the course class in "Introduction to Number Theory" and supervising the thesis.

Hua Loo-Keng said to me many times: "The number theory work of Erdös is just an application of Brun's sieve method." I heard that someone at the Specialization in Probability and Statistics was studying an article by Paul Erdös and others on the estimation of the number of orthogonal Latin squares, so I borrowed the article on Latin squares by D.S. Cowla, Erdös, and A. Strauss to read. They studied the following problem.

A $s \times s$ square formed by s different elements, for example, 1, 2, ..., s, is said to be a Latin square of order s, if every element occurs exactly once in every row and once in every column of the square. Two Latin squares of order s are called orthogonal if, when one of them is superposed on the other, every element of the first square occurs with every element of the second square squares once and only once.

$N(s)$ denotes the maximum number of mutually orthogonal Latin squares of order s. On $N(s)$ Euler had proposed a famous conjectured that

$$N(s) = 0$$

whenever $s \geq 10$ and $s \equiv 2 \pmod 4$.

In 1960, R. C. Bose, S. S. Shrikhande, and E. T. Parker disproved Euler's conjecture. In detail, they proved that if $s > 6$, then

$$N(s) \geq 2.$$

This result caused great concern.

In the same year, Chowla et al., by combining the method of Bose et al. and Brun's sieve method, improved the results of Bose et al. to

$$N(s) > \frac{1}{3} s^{\frac{1}{91}}.$$

This is another good opportunity to go along for the ride! By refining the proof of (3, 4) slightly, I get the estimation: there exists s_0, such that

$$N(s) > s^{\frac{1}{26}}, \quad s > s_0. \tag{5.12}$$

My paper was published in the *Acta Mathematica Sinica*, No. 3, 1966. By this time, the "Cultural Revolution" had begun, and the Acta Mathematica Sinica was ordered to stop publication. This article was just in time to catch the last train.

In 1971, after the "Lin Biao Event", the Institute of Mathematics gradually resumed its research work. By this time, I had entered into the study and research on the Diophantine analysis, hoping to find a way to solve the convergence problem of the high-dimensional numerical integration formula proposed by Hua Loo-Keng and me. I came across J.W.S. Cassels' book *An Introduction to the Geometry of Numbers* (1959), on which there is a theorem of Davenport:

Let Λ be a n-dimensional lattice, and

$$\vec{c}_i (1 \leq i \leq n-1)$$

be any given $n-1$ vectors, then there exists a basis $\vec{a}_i$ $(1 \leq i \leq n)$ of Λ such that

$$|\vec{a}_i - N\vec{c}_i| = O(N^\varepsilon), \quad 1 \leq i \leq n-1$$

holds for any real numbers N (≥ 2) and $\varepsilon > 0$, where the constant in "O" depends only on Λ, ε and $\vec{c}_i$'s.

When I read the proof, I immediately realized that I could use the "sparse sieve" on this problem. In fact, I improved, by using the "sparse sieve", the key part in the proof of the Davenport theorem, i.e., I proved that:

Let $q \geq 2$ be any integer and s, t be any two integers satisfying $(t, q) = 1$. Then there exists an absolute constant K such that there is a positive number u satisfying $(tu + s, q) = 1$ in any interval with length not less than $K \log^3 q$.

So I improved the estimation $O(N^\varepsilon)$ in Davenport's result into

$$O(\log^3 N). \tag{5.13}$$

where the constant in "O" depends only on Λ and $\vec{c}_i$'s.

This was again a by-product work and was completed only in a month. I wrote a short article and submitted it to the *Acta Mathematica Sinica*. At that time, the Cultural Revolution had not yet ended, but the Acta Mathematica Sinica had resumed publication. My article was published in the December 1975 issue. This was the first purely mathematical paper I published after the Cultural Revolution.

5.5 Chinese School of Analytic Number Theory

When I first started working on the sieve method, I attempted to generalize a result obtained by I. V. Chulanovski using the Λ^2-method.

The numbers $0, u_1, ..., u_{m-1}$ are nonnegative integers mutually distinct. U_p denoteing the number of different residues of $\{0, u_1, ..., u_{m-1}\}$ mod p. Assume that $U_p < p$ for any p, and Let $Z_{u_1,...,u_{m-1}}(N)$ denote the number of x such that $x, x + u_1, ..., x + u_{m-1}$ are all prime numbers whenever $x = 1, ..., N$, then

$$\varlimsup_{N\to\infty} Z_{u_1,\ldots,u_{m-1}}(N)\frac{\log^m N}{N} \leq 2^m \cdot m! \prod_p \frac{(1-\frac{U_p}{p})}{(1-\frac{1}{p})^m}.$$

Taking $m=2$ and $u_1=2$, then $Z_2(N)$ is the number of pairs of twin primes not exceeding N. I attempt to generalize the linear function x and $x+u_i(1 \leq i \leq m-1)$ in the above result to the irreducible polynomials. I note that $U_p < p$ holds for all prime numbers p is equivalent to the sufficient and necessary condition for

$$x(x+u_1)\ldots(x+u_{m-1}), \quad x=1, 2, \ldots$$

with no fixed prime factor. Thus I proved that

Let $F(x) = F_1(x)\ldots F_m(x)$ be an integral valued polynomial of order k with no fixed prime factors for all integers x, where $F_i(x)$ ($1 \leq i \leq m$) are irreducible integral valued polynomials mutually distinct. Let $Z(N, F(x))$ denote the number of x such that $F_1(x), \ldots, F_m(x)$ are simultaneously prime when $x=1, 2, \ldots, N$. Then

$$\varlimsup_{N\to\infty} Z(N, F(x))\frac{\log^m N}{N} \leq m!2^m \rho_F. \tag{5.14}$$

here

$$\rho_F = \prod_p \frac{(1-\frac{\omega(p)}{p})}{(1-\frac{1}{p})^m}, \quad \omega(n) = \frac{h_n(k!n)}{k!}, \tag{5.15}$$

where $h_n(k!n)$ represents the number of solutions of the congruence equation $F(x) \equiv 0(\bmod n)(0 \leq x < k!n)$.

Taking $F_1(x) = x$, $F_i(x) = x + u_{i-1}, 2 \leq i \leq m$ gives the Chulanovski theorem stated above.

Familiarize yourself with Hardy and Littlewood's conjecture, by the circle method, about asymptotic formula for the numbers of pair of twin primes not exceeding N. Replacing the right end of (5.14) with c, so that c agrees with Hardy and Littlewood's conjecture about twin prime numbers when $F_1(x) = x$, $F_2(x) = x+2$, then we can conjecture the following asymptotic formula.

$$Z(N, F(x)) \sim c_F \frac{N}{\log^m N} \tag{5.16}$$

At the time, I thought long and hard about whether to include this conjecture in the article or not. Considering that it was only a conjecture and not difficult to guess, I did not include it in the article. Again, since it was not difficult, I submitted the article to *Advances in Mathematics*, which was published in 1957, and it was rather unfortunate to see later that Schinzel and Sierpinski published the conjecture in 1958.

Through the study and research of the sieve method and Goldbach's conjecture, it became perfectly clear to me that for number theory, methodological innovation is

5.5 Chinese School of Analytic Number Theory

much more important than result improvement. I have asked Hua Loo-Keng about this view. He did not deny it, but I could see that he approved of it. Many years later, Hua Loo-Keng once told me, "A whole lot of theory that has no application will not be recognized." That's a much more comprehensive way of looking at things. That's why Hua Loo-Keng always straight into the difficult problems in his delving into number theory. In the process, methods were innovated or existing methods were improved, and his students came to study number theory in the same way. I think this is the characteristic of the so-called Chinese school of analytic number theory.

Chapter 6
Applied Mathematics Explored

Interview

In 1990, you and Professor Hua Loo-Keng's *Applications of Number Theory to Numerical Analysis* (1978) won the Chen Jiageng Prize for Material Science, a recognition of the work of mathematicians by the scientific community and a high honor. This work also had a considerable impact on the international mathematical community and in the applied sector, with no less than 15 book reviews abroad and many more citations.

We would like to know what motivated you to study applied mathematics. And how did you make the "gorgeous transition" from basic mathematics research to applied mathematics research?

6.1 Commitment to Applied Mathematics

I devoted myself to the study and research of applied mathematics for a number of reasons, broadly speaking the following: first, I felt at a loss for number theory. The study of the sieve method and Goldbach's conjecture had no further novelty to develop. Although there later was success in the study of the least primitive root, this was only a short-term act. I was therefore faced with a situation in which I had to change the direction of my research. I have a very narrow base, and if I want to continue my research in number theory, I will first have to read a lot of literature and then have an entry point, and I guess it will be difficult to change my career in a short time. For me, other areas of pure mathematics are even further apart. Secondly, the situation of the "Great Leap Forward" was pressing. The pressure of the so-called "linking theory with practice" was extremely heavy, and the situation forced me to explore the mysteries of applied mathematics. Thirdly, I think the most important thing is that after some investigation, I soon understood some applied mathematics and did not seem to find it difficult to start. In particular, Hua Loo-Keng and I made some results together; found a path that I could continue to study. This led to a keen interest in clarifying some of the problems in further depth. Fourth, Hua Loo-Keng also faced the same pressure of "linking theory with practice", and he pulled me along to go into the field of applied mathematics. This gave me a lot of confidence. Fifth, I had a patriotic passion and a desire to make a direct contribution to nation building.

In 1958, Hua Loo-Keng and I simultaneously probed into the "ore body geometry", "linear programming", and "pseudo-Monte Carlo methods." I never thought that the research and application of pseudo-Monte Carlo methods would stay with me for the rest of my life in the 50 years since then.

6.2 Geometry of the Ore Body

In 1958, Hua Loo-Keng first thought that he should look for mathematical problems in other disciplines. He thought of mining and water resources and many other areas, and for this purpose, he asked me to go to the relevant departments in Beijing to find out what was going on. I went to the mathematics teaching and research section of the colleges of mining, geology, and petroleum as well as other institutions of higher learning and asked them how they were doing mathematics in conjunction with their specialties. They received me very warmly. I borrowed some books on geometry of ore bodies from the mathematics teachers of the College of Mines and the College of Geology. From his friend Lu Shufen, Hua Loo-Keng learned how geographers calculate the area of slopes.

I read the books on geometry of ore bodies and learned practical methods for approximating the volume of an ore body or reservoir, and for approximating the area of a slope. These methods require a "contour map" based on a survey, i.e., a planar cross-section of the object as a basis for calculation. For example, suppose that the map of a slope is a contour map with an elevation difference of $\Delta h = \frac{h}{n}$. It has a commanding point (l_n), its height is h, and contour lines (l_{n-1}), ..., (l_0), they are drawn one by one from the commanding point outward. The difference in elevation between two adjacent contour lines (l_{i+1}) and (l_i) is Δh, and the projected area of the plane is B_i as shown in the following figure.

6.2 Geometry of the Ore Body

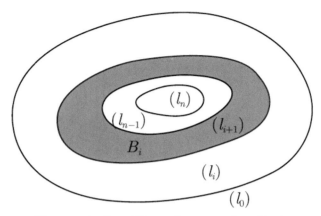

Planar projection of the contour map of a subject

The Bauman method for estimating the area of a slope is described in books on geometry of ore bodies: (i) Find the area of the "upright partition" between two adjacent elevations $C_i = \frac{1}{2}(l_i + l_{i+1})\Delta h$, where l_i denotes length of (l_i). (ii) $\sum_{i=0}^{n-1} \sqrt{B_i^2 + C_i^2}$ is the approximate value of the area of the slope.

Lu Shufen introduced to Hua Loo-Keng the Volkov method used by geographers to calculate the area of a slope: (i) $l = \sum_{i=0}^{n-1} l_i$ is the sum of all the contour lengths, and $B = \sum_{i=0}^{n-1} B_i$ is the projected area of the slope. (ii) $\sqrt{B^2 + (l\,\Delta h)^2}$ is the approximate area of the slope.

Suppose that $\Delta h \to 0$ the area of the slope obtained by the two methods above is B and V, and the true area of the slope is S. What is the relationship between these three? This is not discussed in the books on geometry of ore bodies. Hua Loo-Keng applied the cylindrical coordinate in three dimensions to prove the inequality.

$$V \leq B \leq S$$

That is, both methods yield a slope area that is somewhat smaller than the true area.

Using cylindrical coordinates, I made another study of methods of finding volume from contour map in books on ore body geometry, and got a comparison of the various methods and some new algorithms as well. All those books on ore body geometry are very thick, in which the calculations are quite cumbersome and lack precision due to the language and methods of elementary geometry.

Using a little calculus, we wrote out these methods in only a dozen pages with a precise theoretical analysis that was very clear. We co-authored these results in an article, "On the Calculation of Mineral Reserves and Slope Area on Contour Maps", published in the *Acta Mathematica Sinica*, No. 1, 1961. This was one of the earliest papers on applied mathematics by Hua Loo-Keng and me.

Through this work, although we learned some of the mathematical methods used in mining, surveying, and water resource, and especially wrote a paper, the mathematics involved in the geometry of the ore body was after all relatively simple, and we really could not see what other mathematical problems could be done, and we resolutely ended the work.

6.3 Linear Programming

In 1958, when the Chinese mathematical community started the trends of "linking theory to practice" and "mathematics directly serving the national economy", mathematicians who had long worked in pure mathematics in the traditional way were indeed at a loss. Even mathematicians working in mathematical fields labeled as "linking to practice", such as differential equations, probability, and statistics, were at a loss too because they worked in the same way as mathematicians working in pure mathematics.

At the initiative of Tsien Hsue-Shen, the Institute of Mechanics set up an operations research unit. They combined linear programming, quality control, input–output methods, etc., and did theoretical work while looking for application problems in the industrial sector, also known as "mission work". They wrote a booklet to popularize operations research. At this time, "linear programming" was like a lightning bolt that opened the eyes of the Chinese mathematical community. It was amazing that mathematics could directly serve the national economy.

I read their booklet and quickly understood that linear programming is the problem of finding the global extremum of a linear objective function $f(\vec{x})$ on a high-dimensional convex polyhedron D defined by a linear inequality. It is known from calculus that the global extremum of $f(\vec{x})$ cannot generally be taken at the interior points of D, but it can certainly be reached at the boundary ∂D of D, since $f(\vec{x})$ is still linear on ∂D, so by analogy, we get the conclusion that the point that makes the objective function reach the global extreme value on D must be reached at some vertex of D. The so-called "linear programming" is a method of approximating the vertices of D that will lead $f(\vec{x})$ to the global extremum.

At that time, one of the background problems of linear programming was the transportation of grain: suppose there is a map and a grain to be transported, and there are n cities $A_1, ..., A_n$ on the road of the map to send $a_1, ..., a_n$ tons of grain, and m cities $B_1, ..., B_m$ on the road, each of which has to bring in $b_1, ..., b_m$ tons of grain. We know the distance between any two cities and assume $\sum_{i=1}^{n} a_i = \sum_{j=1}^{m} b_j$, and ask how can we develop a dispatch plan that minimizes the total cost.

At that time, "theory linked to practice," "plucking white flag," and "pompous wind" swept the national science and technology and education circles. Hua Loo-Keng stepped aside. The research divisions of the Institute of Mathematics were abolished and replaced by four so-called "commanding units" set up according to the army establishment, each with a director and a political commissar. Probability and Statistics, Differential Equations and Functional Analysis, Mathematical Logic, and

6.3 Linear Programming

Computer Logic Design were each established as a command unit. Those who were working in several abstract disciplines, such as number theory, algebra, topology, etc., in the Institute of Mathematics found their own way out. Some of them moved to other disciplines, while others formed a command called "Operations Research". Wan Zhexian was the director and Zhu Yongjin was the political commissar. At the beginning, Wan Zhexian was very active, but soon he stopped being active at all. Later on, this command was merged with the Operations Research Division of the Institute of Mechanics to form the Operations Research Division of the Institute of Mathematics, and this was an afterthought.

Xu Kongshi and Wei Daozheng from our number theory group went to work at the Institute of Computing Technology of the Chinese Academy of Sciences long ago; Yue Minyi and Wu Fang joined the Operations Research Division and have stayed there ever since; I also joined the Operations Research Division. Hua Loo-Keng was the director of the Institute of Computing Technology of the CAS when it was organized. I spent some time figuring out linear programming. Others, such as queuing theory and game theory, I asked very little about. At this time, I was still following the progress of number theory and other approaches of applied mathematics.

During the "Great Leap Forward", the Institute of Mathematics enrolled more than a hundred teachers from higher education institutions in China for further training. Most of them came from teacher training colleges and engineering colleges. They were politically selected. Generally speaking, they were relatively poor in mathematics and had not done much research work before. Therefore, there was no such "baggage" as "research direction". In the "Great Leap Forward", they were even more "left". They were more receptive to the "linking theory to practice" and "breaking the old system." The leadership of the four commands and other social affairs at the Mathematical Institute was then mostly in their hands. It was not until after 1961 that they gradually returned to their original institutions.

I had a good personal relationship with those visiting teachers in the operations research division. Since I had a good foundation in mathematics and some experience in doing research, I was able to grasp the algorithmic theory of linear programming relatively quickly. I co-authored a pamphlet with Wan Zhexian, *Mathematical Methods in Material Transport* (Science Press, 1959), and a textbook with Zhu Yongjin and others, *Theory of Linear Programming and Its Applications* (Higher Education Press, 1959). Both of these books were published under the name of the Operations Research division of the Institute of Mathematics, Chinese Academy of Sciences. Their publication had a positive effect on the training those visiting teachers and the popularization of linear programming in China. I have often given lectures to those teachers for training or to the "Short Course on Operations Research" organized by the Operations Research Division. I learned that most of the teachers were interested in learning more mathematics and really saw me as a teacher. We will always remember this time together. It was not until 2005, almost 50 years later, when I went to Chongqing on a business trip, that Tan Yuqing, one of the visiting teachers of that year, brought her daughter and grandson to visit me at the hotel. In the same year, when I visited Southwest Normal University, Huang Shaowen came to see me immediately. They were already the backbone of the local teachers.

The Institute of Mathematics was open to the whole country and receiving teachers from other universities and colleges for further training, which should have been an achievement at that period. By the 1980s, when I was the director of the Institute of Mathematics, the first thing that came to my mind, together with the deputy director, Yang Lo, was to open up the Institute of Mathematics and to receive visiting scholars. By this time, China had opened up to the world, and the level of mathematics in the country had improved. Relatively speaking, the status of the Institute of Mathematics was not as high and important as it was then.

Through this work, I learned about linear programming and participated in writing and lecturing on two booklets. At the same time, I knew that the mathematics used here was simple and I did not see any further mathematical problems to work on, especially since my expertise in number theory was not used. Operations research had to be combined with "mission work", and I was unwilling and afraid to stray too far from mathematics, so I decided to end the work.

6.4 Pseudo-Monte Carlo Method

In 1958, attention was drawn to the summary of the work of the USSR Academy of Sciences for 1957, in which two important mathematical results were mentioned: one on mathematical methods in public utilities, the "theory of queuing", and the other on "Applications of number theory to the approximation of multiple integrals." According to Hua Loo-Keng's recollection, Wang Yuan took an article to Hua Loo-Keng to talk about the Monte Carlo method in the Numerical calculation of integrals, and the article pointed out that the random numbers needed to calculate the integrals obeyed a uniform distribution, and etc. In my memory, I did take I.M. Korobov's article to go to Hua Loo-Keng. That day, Hua Loo-Keng was very tired and did not want to read it, so I said, "Just this line, okay?" Hua Loo-Keng got excited after reading it and said, "The Monte Carlo method is actually an application of uniform distribution theory in number theory. It's like poking a hole through a window, spacing out and poking through just a little something."

Anyway, in 1958 we finally got in the Soviet "Scientific Record" to the article by N. M. Korobov on the application of number theory to the approximation of multiple integrals, i.e., the work mentioned in the summary of the work of the USSR Academy of Sciences in 1957.

The article is simple and easy to understand. Assume $f(\vec{x})$ is a periodic function in unit cube

$$U_s : 0 \leq x_1, ..., x_s \leq 1$$

with period 1 for each variable, and it has the Fourier expansion.

$$f(\vec{x}) = f(x_1, ..., x_s) = \sum \cdots \sum_{-\infty}^{\infty} C(m_1, ..., m_s) e^{2\pi i(m_1 x_1 + ... + m_s x_s)},$$

6.4 Pseudo-Monte Carlo Method

here the Fourier coefficients satisfy

$$|C(m_1, ..., m_s)| \leq C/(\overline{m}_1, ...\overline{m}_s)^\alpha,$$

where $C > 0$ and $\alpha > 1$ are absolute constants, $\overline{m} = \max(1, |m|)$. The class of functions formed by the entirety of such functions is denoted as $E_s^\alpha(C)$. Learn to know

$$C(0, ..., 0) = \int_{U_s} f(\vec{x})d\vec{x}, \quad d\vec{x} = dx_1...dx_s, \tag{6.1}$$

let p to be a prime, Korobov suggests using the single sum

$$\frac{1}{p^2} \sum_{k=1}^{p^2} f\left(\frac{k}{p^2}, \frac{k^2}{p^2}, ..., \frac{k^s}{p^2}\right) \tag{6.2}$$

to approximate the multiple integral (6.1). Substituting the Fourier expression of $f(\vec{x})$ for (6.2) and making the difference between (6.1) and (6.2) yields

$$\Delta = \int_{U_s} f(\vec{x})d\vec{x} - \frac{1}{p^2} \sum_{k=1}^{p^2} f\left(\frac{k}{p^2}, \frac{k^2}{p^2}, ..., \frac{k^s}{p^2}\right)$$

$$= \frac{1}{p^2} \sum_{k=1}^{p^2} \sum_{\vec{m}}' C(m_1, ..., m_s) e^{2\pi i(m_1 k + ... + m_s k^s)/p^2},$$

here $\sum_{\vec{m}}'$ denotes the omitting of the term with $m = \vec{0} = (0, ..., 0)$, so that for all $f \in E_s^\alpha(C)$, the upper bound on the error is

$$|\Delta| \leq \sum_{\vec{m}}' \frac{C}{(\overline{m}_1, ..., \overline{m}_s)^\alpha} \left|\frac{1}{p^2} \sum_{k=1}^{p^2} e^{2\pi i(m_1 k + ... + m_s k^s)/p^2}\right| \tag{6.3}$$

Thus, the estimate of the error (6.3) boils down to an estimate of the complete trigonometric sum at the right end of (6.3). Such estimate of the complete trigonometric sum has already been done by Hua Loo-Keng. Here, Korobov takes the number of terms of the trigonometric sum to be the square of the prime p so that a particularly precise estimate is obtained:

$$\frac{1}{p^2} \sum_{k=1}^{p^2} e^{2\pi i(m_1 k + ... + m_s k^s)/p^2} = O(p^{-1}).$$

This leads to

$$|\Delta|=O(p^{-1}).$$

The principal order of his estimate is independent of the dimension s of the integral. This is a very surprising result.

In 1959, we meet Korobov's next article, in which he uses

$$\frac{1}{p}\sum_{k=1}^{p} f\left(\frac{a_1 k}{p}, ..., \frac{a_s k}{p}\right) \qquad (6.4)$$

to approximate (6.1), where $\vec{a}=(a_1, ..., a_s)$ is an integer vector that depends only on the p. He proves that there exists an integer vector $\vec{a}$ such that the error of the approximation does not exceed.

$$O(\log^{\alpha s} p / p^{\alpha}). \qquad (6.5)$$

From the point of view of computational mathematics, his results cannot be considered "constructive", so the desire to obtain a constructive method becomes the central issue.

In fact, the upper bound on the difference between (6.1) and (6.4) is

$$\left| \int_{U_s} f(\vec{x})d\vec{x} - \frac{1}{p}\sum_{k=1}^{p} f(\frac{a_1 k}{p}, ..., \frac{a_s k}{p}) \right|$$
$$\leq \sum_{a_1 m_1 + ... + a_s m_s \equiv 0(\bmod p)}{}' \frac{C}{(\overline{m}_1, ..., \overline{m}_s)^{\alpha}} = \Delta'.$$

So Δ' to be small, the whole vector $\vec{a}=(a_1, ..., a_s)$ must be taken so that the congruence

$$a_1 m_1 + ... + a_s m_s \equiv 0 (\bmod p)$$

has only one trivial solution $\vec{m} = \vec{0}$ in an area as large as possible

$$\overline{m}_1, ..., \overline{m}_s \leq M$$

i.e., requiring M as large as possible.

Hua Loo-Keng, with his characteristic strong mathematical intuition and his straightforward approach of starting with concrete examples, suggested that I should first try the problem of estimating the lower bound of Δ' by means of the continuous fractions. With respect to the lower bound of Δ', we obviously have $\Delta' \geq C/p^{\alpha}$. I proved $\Delta' \geq C c \log p / p^{\alpha}$ that I got a slight advantage by using continuous fractions. Hua Loo-Keng further suggested that I try the upper bound estimate for the double integral by using the golden number $\frac{\sqrt{5}-1}{2}$ and the Fibonacci sequence. At that time, my knowledge of continuous fractions was only a little Learned from the *Introduction to Number Theory*. Since $\frac{\sqrt{5}-1}{2}$ is a number with a denominator of 2, it seemed inconvenient. I then switched to the unit $\sqrt{2}-1$ in $Q\sqrt{2}$, thus proving a two-dimensional integration formula

$$\int_0^1 \int_0^1 f(x_1, x_2) dx_1 dx_2 - \frac{1}{q} \sum_{k=1}^q f\left(\frac{a_1 k}{q}, \frac{a_2 k}{q}\right) = O\left(\frac{\log q}{q^\alpha}\right), \qquad (6.6)$$

where $f \in E_2^\alpha(C), a_1 = 1, a_2 = p_n = \frac{1}{2}\left((1+\sqrt{2})^n + (1-\sqrt{2})^n\right)$, and $q = q_n = \frac{1}{2\sqrt{2}}\left((1+\sqrt{2})^n - (1-\sqrt{2})^n\right)$. This improves (6.5). The error order of this formula is perfect and constructible from the lower bound estimate, i.e., here

$$\lim_{n \to \infty} \frac{p_n}{q_n} = \sqrt{2}.$$

If we follow the original suggestion of Hua Loo-Keng, then $a_1 = 1$, $a_2 = p_m = F_{m-1}$, $q = q_m = F_m$, where $\{F_m\}$ is the Fibonacci sequence, i.e., the sequence of integers that fits the recursive formula $F_0=0$, $F_1=1, F_m = F_{m-1} + F_{m-2} (m \geq 2)$. At this point,

$$\lim_{m \to \infty} \frac{F_{m-1}}{F_m} = \frac{\sqrt{5}-1}{2} = 2\cos\frac{2\pi}{5}.$$

Hua Loo-Keng came up with the idea of generalizing two dimensions to higher dimensions by constructing an integer vector $\vec{a}=(a_1, ..., a_s)$ mod n in terms of a set of independent units in cyclotomic field $Q(2\cos\frac{2\pi}{p})$. That idea kept me working for years to fully figure out.

We co-authored a paper "Remarks concerning Numerical integration" on the two-dimensional integration formula (6.6) together with the lower bound estimate for numerical integrals, published in the *Science Record*, New Ser. Vol.4, No.1, 1960. Later, we know that the similarity of the formula (6.6), constructed in terms of the Fibonacci sequence, had also been proved independently by N. S. Bakhvalov. The method of our proof is different.

6.5 The Hua-Wang Method

In fact, there is a close relationship between the Fibonacci sequence and the golden number $\frac{\sqrt{5}-1}{2}$, i.e.,

$$\left| \frac{F_{n-1}}{F_n} - \frac{\sqrt{5}-1}{2} \right| < \frac{1}{\sqrt{5}F_n^2}$$

(see Chap. 10 of Hua Loo-Keng's *Introduction to Number Theory*). Further, we have

$$\left(\frac{\sqrt{5}+1}{2} \right)^n = F_{n-1} + F_n \left(\frac{\sqrt{5}+1}{2} \right). \ n = 1, 2, \ldots \quad (6.7)$$

Since $\frac{\sqrt{5}+1}{2}$ is the unit of a quadratic field $Q(\sqrt{5})$ and $\left(1, \frac{\sqrt{5}+1}{2}\right)$ is a set of bases of $Q(\sqrt{5})$, a two-dimensional pseudo-random series is

$$\left(\frac{k}{F_n}, \left\{ \frac{F_{n-1}k}{F_n} \right\} \right), \ k = 1, \ldots, F_n, \ n = 2, 3, \ldots$$

where $\{x\}$ denotes the fractional part of x. Considering $\frac{\sqrt{5}-1}{2} = 2\cos\frac{2\pi}{5}$, Hua Loo-Keng proposed to extend the field $Q(\sqrt{5})$ to $Q(2\cos\frac{2\pi}{p})$, where p is a prime number ≥ 7. This is a totally real algebraic number field of the degree $s = \frac{p-1}{2}(\geq 3)$. We know that it has a set of independent units $\varepsilon_1, \ldots, \varepsilon_{s-1}$, and a set of integer bases

$$\omega_j = 2\cos\frac{2\pi j}{p}, \ j = 1, \ldots, s$$

From $\varepsilon_1, \ldots, \varepsilon_{s-1}$, we can construct a unit system $\eta_1 < \eta_2 < \ldots$ that satisfies

$$\eta_j^{(i)} \ll \eta_j^{-\frac{1}{s-1}}, \ i = 2, \ldots, s \quad (6.8)$$

where $\eta_j^{(1)} = \eta_j$, $\eta_j^{(i)}(2 \leq i \leq s)$ is the conjugate number of η_j. If $s = 2$, then $\eta_j = \eta_j^{(1)} = \left(\frac{\sqrt{5}+1}{2}\right)^j$, $\eta_j^{(2)} = \left(\frac{\sqrt{5}-1}{2}\right)^j$. Imitating (6.7) for $s = 2$ and expressing η_j in terms of the integral bases yields

$$\eta_j = n_1\omega_1 + \ldots + n_s\omega_s,$$

we can obtain the s-dimensional pseudo-random sequence

$$\left(\frac{k}{n_1}, \left\{ \frac{n_2 k}{n_1} \right\}, \ldots, \left\{ \frac{n_s k}{n_1} \right\} \right), \ k = 1, 2, \ldots, n_1,$$

This was the block diagram for constructing a high-dimensional pseudo-random sequence as proposed by Hua Loo-Keng at that time. For a period of time, I went to Hua Loo-Keng's house every day and shared breakfast with him. After the meal, we worked together on how to estimate the error of the integration formula constructed

6.5 The Hua-Wang Method

with this set of pseudo-random sequences. We could not get a theoretical estimate of the error, so I came up with the idea that with the error given by

$$\lim_{m \to \infty} \frac{F_{m-1}}{F_m} = \frac{\sqrt{5} - 1}{2},$$

$n_1, ..., n_s$ then should have the following properties:

$$\lim_{n_1 \to \infty} \frac{n_j}{n_1} = \omega_j, \; j = 2, ..., s$$

I have modified the above framework to make $n_1, ..., n_s$ to have the above properties. The final form obtained after simplification is: let

$$n_j = \sum_{l=1}^{s} \eta_j^{(l)}, \; h_j^{(i)} = \sum_{l=1}^{s} \omega_i^{(l)} \eta_j^{(l)}, \; i = 2, ..., s$$

then n_j and $h_j^{(i)}$ are both rational integers, and it follows immediately from (6.8) that

$$\left| \frac{h_j^{(i)}}{n_j} - \omega_i \right| = O\left(n_j^{-1-\frac{1}{s-1}} \right), \; i = 2, ..., s$$

Thus a pseudo-random sequence is obtained

$$\left(\frac{k}{n_j}, \left\{ \frac{h_j^{(2)} k}{n_j} \right\}, ..., \left\{ \frac{h_j^{(s)} k}{n_j} \right\} \right), \; k = 1, 2, ..., n_j \tag{6.9}$$

When I was studying this topic, my knowledge of algebraic number theory was only Chapter 16 of Hua Loo-Keng's *Introduction to Number Theory*. It was this chapter that I found the most difficult to understand. I had only a formal understanding of some concepts. The conjugate numbers, conjugate fields, units, etc., were both unfamiliar to me, let alone mastered and applied. The units of an algebraic number field form a finite generated abelian group (Dirichlet's theorem), but this is only an existence theorem. In applied mathematics, we need constructive results, i.e., concrete expressions of the units were required. This pushed me to study algebraic number theory in more depth, learning as I went along. We did not yet know examples of basic unit expressions other than quadratic fields. But we were fortunate to find in David Hilbert's *Report on Number Theory* a set of independent units for the cyclotomic field $Q(2 \cos \frac{2\pi}{p})$, i.e., a set of units with non-zero adjusters. Using it, we construct a set of units that fits (6.8), which gives us the pseudo-random sequence (6.9). This allows us to derive the integration formula

$$\int_{U_s} f(\vec{x})d\vec{x} \approx \frac{1}{n_j} \sum_{k=1}^{n_j} f\left(\frac{k}{n_j}, \left\{\frac{h_j^{(2)} k}{n_j}\right\}, ..., \left\{\frac{h_j^{(s)} k}{n_j}\right\}\right). \qquad (6.10)$$

We could not yet estimate the error in (6.10).

We know that in the pseudo-random series obtained by Korobov's method

$$\left(\left\{\frac{a_1 k}{p}\right\}, ..., \left\{\frac{a_s k}{p}\right\}\right), \quad k = 1, 2, ..., p$$

the amount of the elementary operations required for an integer vector $\vec{a}=(a_1, ..., a_s)$ is $O(p^2)$, and for modulo $p_1 p_2$ it can be reduced to $O((p_1 p_2)^{4/3})$. Thus, Korobov only computed some pseudo-random series for $s = 3, ..., 10$, and obtaining the series (6.9) requires only $O(\log n_j)$ elementary operations. I worked out a pseudo-random series in 11-dimensional space on a mechanical desk calculator at the Institute of Mathematics

$$\left(\left\{\frac{a_1 k}{q}\right\}, ..., \left\{\frac{a_{11} k}{q}\right\}\right), \quad k = 1, 2, ..., 9 \qquad (6.11)$$

where $q = 698047$, $a_1 = 1$, $a_2 = 685041$, $a_3 = 646274$, $a_4 = 582461$, $a_5 = 494796$, $a_6 = 384914$, $a_7 = 254860$, $a_8 = 107051$, $a_9 = 642292$, $a_{10} = 467527$, $a_{11} = 284044$.

I asked my friends at the Institute of Computing Technology to calculate the upper bound on the error of the integration formula corresponding to the series (6.11). Hua Loo-Keng and I wrote a paper on the above result and published it in *Scientia Sinica*, Vol.14, No. 7, 1965. In 1966, when the "Cultural Revolution" began, the *Scientia Sinica* was ordered to stop publication. We finally caught the "last train".

During the Cultural Revolution, we had no idea how our work would be reflected internationally. In 1972, Hua Loo-Keng was part of a delegation to Japan headed by Chengzhi Liao, and he learned that our method was already known to his colleagues abroad. S. Haber first discussed this method in an article and called it "Hua-Wang Method".

How can the discrepancy of the pseudo-random series (6.9) and the error in the integration formula (6.10) be estimated? I have foreseen the need to look for tools in algebraic number theory, Diophantine analysis, and the geometry of numbers. This has greatly stimulated my motivation to study these areas. I browsed extensively in these areas and read a handful of classics such as E. Hecke's *Lectures on Theory of Algebraic Number* and Cassels' *An Introduction to Diophantine Approximation*. In the library, I happened to come across W.M. Schmidt's article in *Acta Math* on the simultaneous approximation of algebraic numbers, which could just be used to solve our problem. Thus, I obtained the deviation of (6.9) and the error estimate of (6.10).

Although the Cultural Revolution had not yet ended, some academic journals resumed publication after the "September 13 Event". Hua Loo-Keng and I published

6.5 The Hua-Wang Method

the above results in an article in *Scientia Sinica* (see 1973, pp. 339–357; 1974, pp. 304–322).

The cyclotomic field method consists mainly in finding a sequence of units $\eta_1 < \eta_2 < \ldots$ satisfying (6.8). If we can find a set of algebraic integers $\alpha_1 < \alpha_2 < \ldots, \alpha_j$ which is itself large and whose conjugates are all small, then we can use $\{\alpha_j\}$ instead of $\{\eta_j\}$. With this idea in mind, Hua Loo-Keng and I wanted to find tools from the generalization of continuous fractions. I found J. Bernstein's book, *The Jacobi-Perron Algorithm—Its Theory and Application*. This is a generalization of continuous fractions. We also found in literature the definition of PV (Pisot-Vijayaraghavan) numbers.

Assuming α satisfy the irreducible equations of integer coefficients

$$f(x) = x^s - a_{s-1}x^{s-1} - \ldots - a_1 x - a_0 = 0,$$

and

$$\alpha(=\alpha^{(1)}) > 1, \ |\alpha^{(2)}| \leq \ldots \leq |\alpha^{(s)}| \leq 1,$$

α is said to be a PV number.

Let $\alpha_j = \alpha^j$ ($j = 1, 2, \ldots$). We substitute α_j for η_j, then the simultaneous rational approximation of α_j^k ($1 \leq k \leq s-1$) follows from the approach of the previous section

$$\left| \frac{S_{n+k}}{S_n} - \alpha_j^k \right| \ll S_n^{-1-\rho}, 1 \leq k \leq s-1 \tag{6.12}$$

where $\rho = -\log \left|\alpha_j^{(s)}\right| \Big/ \log \alpha^j$ and $S_l = \sum_{i=1}^{s} \alpha^{(i)l}$ are rational integers.

To have the smaller right end of (6.12), i.e., ρ the larger, one must find an equation $f(x) = 0$, which has the conjugates of all α about the same in absolute value in addition to one larger real root α, then it would be possible to substitute $\{\alpha^j\}$ for $\{\eta_j\}$ in the cyclotomic field method. The advantage of this is that the sum of the conjugates of α^j can be reduced to a primary symmetric function, which can be calculated by Newton's formula. So the calculation is simpler. However, except the case of $s = 3$, PV number α cannot be found making $|\alpha^{(i)}| = \alpha^{-\frac{1}{s-1}}, 2 \leq i \leq s$, so the accuracy is always less than that by the cyclotomic field method.

Thus, the key problem is to find a good PV number. We first thought of a generalization of the golden number, i.e., consider the equation

$$F(x) = x^s - x^{s-1} - \ldots - x - 1 = 0.$$

Note here that the root of $F(x) = 0$ is right $\frac{1 \pm \sqrt{5}}{2}$ when $s = 2$.

We can show that $F(x)$ is an irreducible polynomial and the maximal real root η of $F(x) = 0$ and its conjugate number satisfying

$$2 - 2^{-(s-1)} < \eta < 2 - 2^{-s},$$

together with

$$\left|\eta^{(i)}\right| \leq \eta - 1, \ 2 \leq i \leq s$$

So η is a PV number.
Let

$$F_0 = \ldots = F_{s-2} = 0, \ F_{s-1} = 1, \ F_{n+s} = F_{n+s-1} + F_{n+s-2} + \ldots + F_n, \ n = 0, 1, \ldots$$

$Q(\eta)$ has a set of integer bases

$$\omega_l = \eta^{l-1} - \ldots - \eta - 1, \ 1 \leq l \leq s.$$

We obtain

$$\left|\frac{F_n(j)}{F_n} - \omega_j\right| = O\left(F_n^{-1 - \frac{1}{2^s \log 2} - \frac{1}{2^{2s+1}}}\right), \ 2 \leq j \leq s,$$

where $F_n(j) = F_{n+j-1} - F_{n+j-2} - \ldots - F_n, 2 \leq j \leq s$. Thus, the pseudo-random series is obtained as follows:

$$\left(\frac{k}{F_n}, \left\{\frac{F_n(2)k}{F_n}\right\}, \ldots, \left\{\frac{F_n(s)k}{F_n}\right\}\right), \ 2 \leq k \leq F_n$$

We can also consider irreducible polynomials

$$G(x) = x^s - Lx^{s-1} - 1,$$

that is better than the result obtained with $F(x)$ when L is larger.

Hua Loo-Keng and I wrote up the results in this regard in an article published in *Sientia Sinica*, Vol.18, No. 2, 1975.

By this time, the Institute had more advanced computers, and Rongxiao Zhang used the above methods to work out some pseudo-random series for $12 \leq s \leq 18$.

6.6 Book Review of *Applications of Number Theory to Approximate Analysis*

As early as 1961, Hua Loo-Keng had co-authored a booklet with me, *Numerical Calculation of Integral* (Science Press, 1961). Two years later, a revised version of the book, *Numerical Integration and Its Applications*, was published. I also wrote

6.6 Book Review of *Applications of Number Theory to Approximate Analysis* 63

a review paper, *On Numerical Integration and Its Applications* (Progress in Mathematics, 1962, 1–44). None of these works included the main works of Hua Loo-Keng and me, namely, the method of cyclotomic field and the method of PV numbers. After the end of the 'Cultural Revolution', we wrote a monograph, *Applications of Number Theory to Numerical Analysis* (Science Press, 1978), which provides a more comprehensive summary of the subject. Science Press and Springer Publishers wanted to jointly publish in English a number of outstanding mathematical monographs in China. This book is the first in the series. Since the book was published in 1981, it has seen more than a dozen reviews. Since this book was the first more comprehensive summary of the field, it has often been used as a basic reference for subsequent works in the field. One review of the book is now cited below.

What is number theory good for? No one doubts that many branches of mathematics owe their existence to—or, at least, were strongly stimulated by—problems of the 'real world', like those of physics, engineering, etc. Familiar examples are the calculus and the theory of differential equations needed in celestial mechanics; partial differential equations that are indispensable in hydrodynamics and so on. But number theory? Often number theorists, when challenged by our first question (usually asked by nonmathematicians) feel obligated to convince the questioner that number theory also can be useful. Sometimes its applications in problems of crystallography and, more recently, in cryptography are mentioned. Why it should be necessary to point out a 'usefulness' in the commonly understood sense for number theory is something of a mystery to this reviewer. It appears quite certain that Diophantus, or Fermât, or Gauss studied this field of human knowledge because of its intrinsic interest and its peculiar beauty—and they really did not care one way, or the other, whether their elegant theorems would, or would not have 'useful' applications. Be that as it may, it turns out that like so many other branches of mathematics, developed by the 'purest' of mathematicians, also number theory does have applications outside itself. In addition to cryptography, or to the numerous problems of physics concerned with lattices (crystallography is just one of them; the study of perfect gases is another one; see [2]), one may list the many applications of number theory to the theory of computers (see, e.g., [9, Vol. 2]), the generation of random numbers (see [15, or 4]), and many more. Two relatively recent publications, [23] edited by Zaremba and the present book by Hua and Wang, called attention to yet another great field of number theoretic application, namely numerical analysis. At present, the number theorist who feels compelled to justify to the world the love for his field by the latter's " usefulness" can proudly point to the need for sophisticated number theory not only in [23] and in Hua and Wang but also Knuth's Art of computer programming [9], numerous papers by Dieter (see, e.g., [23, pp. 287–317] and [5]), and many more. While Zaremba's collection of papers [23] discusses applications of number theory to numerical analysis understood in a rather broad sense, the book by Hua and Wang concentrates upon essentially one problem only, namely the numerical computation of multiple integrals. Indeed eight of its ten chapters are devoted to it, and only the last two chapters discuss other topics (interpolation and differential and integral equations). Numerical integration is almost as old as integration itself. Some of the classical polynomial interpolation formulae are due to Newton. In fact,

Newton himself (already in 1676; see, e.g., [10, p. 231]) used those formulae for the approximate computation of definite integrals. Better known and more accurate formulae for "mechanical quadrature", as numerical integration (especially in one variable) used to be called until quite recently (see, e.g., [10 and 16]), are those of Simpson, Weddle, Stirling, Bessel, Lagrange, and Gauss, among others. While originally devised only for simple integrals, they were soon used, by iteration, also for double integrals. Nevertheless, number theory played hardly any role in the most important problem related to these formulae, namely the estimation of the maximal error term; this was obtained by analytic methods.

The main idea of the quadrature formulae is to replace an integral $\int_a^b f(x)dx$ by a finite sum $\sum_{n=0}^k a_n f(x_n)$, $a = x_0 < x_1 < ... < x_k = b$, a_n an independent of f, in such a way that the size of the error term $R_k(f) = \left| \int_a^b f(x)dx - \sum_{n=0}^k a_n f(x_n) \right|$ should be as small as possible, for any function f of a given class (such as functions of bounded variation, continuous functions, twice differentiable functions, etc.).

In many of the formulae, the points x_n are obtained by simply subdividing the interval [a, b] into k equal parts. However, in Gauss' formula, normalized for the interval $[-1, +1]$, the x_n are the zeros of the nth Legendre polynomial. Gauss showed that, for the same amount of computation (as measured, e.g., by the number and precision of the terms used) one can improve considerably the result by a judicious choice of the points of subdivision.

This remark may be considered as the starting point of many future developments. Indeed, the question arises: can we select the points where we compute the function to be integrated in such a way that the average of very few such functional values, perhaps with proper weights, should yield the value of the integral over a unit interval with a minimal error? And next, can this procedure be used in a Euclidean space of arbitrarily many dimensions? This second question is far from trivial. Indeed, any integral in one variable, $\int_a^b f(x)dx$ can be normalized to an integral over the unit interval [0,1] by a simple linear change of the variable. This, however, is no longer the case even in two dimensions; not every simply closed curve, even if convex, can be mapped onto the unit square by similarity transformations and rigid motions. The problem is still more difficult in higher dimensions and, in particular, for nonconvex volumes. Even in these cases, however, the very definition of the Riemann integral suggests that if we average the value of a reasonably smooth function at sufficiently many points with a sufficiently regular distribution over a unit volume of fairly arbitrary shape, we ought to obtain a good approximation to the integral of that function over the given volume. With the advent of electronic computers, it became possible to implement this approach. An element of randomness entered the choice of points, and that earned the method the name of Monte Carlo. The Monte Carlo method knew a number of successes, but soon also its limitations became apparent. It became clear once more that one could improve the result by a judicious choice of points. This was realized by Korobov (see [11, 12 and 13]) and, independently and almost simultaneously, by Hlawka [8] somewhat over 20 years ago. These events signaled the start of the thorough use of number theory in numerical analysis. Shortly afterward, again independently, a similar method was used by Conroy (see [3]) in the

6.6 Book Review of *Applications of Number Theory to Approximate Analysis*

evaluation of a multiple integral that occurs in physical chemistry. Today this method, originally called by Korobov the optimal coefficients method, is usually referred to as the good lattice points (g.l.p.) method. The leading ideas of the method are as follows: One attempts to find a good approximation to $\int_0^1 \ldots \int_0^1 f(\vec{x})d\vec{x}$ by a finite sum, as before. Here $\vec{x} = (x_1, \ldots, x_s)$ is a vector in s-dimensional Euclidean space and $d\vec{x} = dx_1 \ldots dx_s$. The accuracy of the procedure is much improved if one assumes that $f(x)$ is periodic, of period one, in each of its s variables. This means that it possesses a multiple Fourier series $f(\vec{x}) = \sum_{\vec{m}} C(\vec{m}) e^{2\pi i (\vec{x}, \vec{m})}$, where $\vec{m}$ runs through all integral vectors in s-dimensional space and $(\vec{x}, \vec{m}) = \sum_{i=1}^s x_i m_i$ is the inner product. Under these conditions we consider sums of the form $n^{-1} \sum_{r=1}^n f(r\vec{\alpha}/n)$. Korobov and Hlawka have shown that it is possible to choose the vector $\vec{\alpha} = \vec{\alpha}(n)$ in such a way that for all functions $f(x)$ of a certain class

$$\left| \int_0^1 \ldots \int_0^1 f(\vec{x})d\vec{x} - \frac{1}{n} \sum_{r=1}^n f(\frac{r}{n}\vec{\alpha}) \right| \leq C n^{-\alpha} (\log n)^\beta.$$

Here C is an absolute constant (for the given class of functions) and α depends on the smoothness of those functions. The vector $\vec{\alpha} = \vec{\alpha}(n)$ has to be chosen by some (not uniquely determined) method, so that the Diophantine equation

$$(\vec{\alpha}, \vec{m}) \equiv 0 \ (\text{mod} \, n), \quad \vec{m} \neq \vec{0} \qquad (6.13)$$

should have no "small" solutions $\vec{m} = (m_1, \ldots, m_s)$. Specifically if we set $\overline{m}_i = \max(1, |m_i|)$, one requires that $\|\vec{m}\| = \prod_{i=1}^s \overline{m}_i$ should exceed a certain lower bound for all solutions $\vec{m}$ of (6.13) (see, e.g., Lemma 3.9). Once we settle on a method for the selection of $\vec{\alpha}$ as a function of n, we can, using the value of $\vec{\alpha}$, also computeβ. It is clear that by taking n successively larger we can reduce the error below any preassigned limit, but for each new n, $\vec{\alpha} = \vec{\alpha}(n)$ has to be recomputed—and that is by far the most time-consuming part of the process. According to the method used to determine $\vec{\alpha}$, the authors speak of *p*-points, good points (g.p.) and good lattice points (g.l.p.). For the purpose of this review we shall largely ignore these distinctions.

Although there exists a theorem to the effect that the measure of the set of good points is one, the effective construction of even a single one of them is not trivial (see Baker [1 and Schmidt 17, 18]). Korobov, Zaremba [19, 21], Halton [7] and others have suggested methods for the economical construction of g.l.p., but one of the most important contributions to this topic is due to the present authors (who, by the way, define g.l.p. in a way slightly different from their predecessors).

Once this problem is solved, one has to extend the results (a) to functions that fail to be periodic with period 1, and (b) to volumes other than s-dimensional cubes. At least three methods have been proposed to handle problem (a). The simplest is probably to set

$$F(x) = \frac{1}{2}\{f(x) + f(1-x)\}.$$

Then $F(0) = F(1), F(x)$ can be defined outside [0,1] by periodicity as a continuous function. It also is clear that

$$\int_0^1 f(x)dx = \int_0^1 F(x)dx,$$

and the previous method applies to the last integral. The generalization to s variables is, of course, immediate. Other methods that have been proposed to overcome the lack of periodicity are changes of variables and the use of Bernoulli polynomials (besides the present book also see, e.g., [6]). As for (b), the solutions suggested are perhaps not entirely satisfactory. An obvious approach is the introduction of the characteristic function for the set S: if $x \in S, \chi(x) = 1$ otherwise $\chi(x) = 0$.

Then, if $S \subset G_s$ (G_s=s-dimensional unit cube), we may apply previous methods to $F(x) = \chi(x)f(x)$ and integrate over Gs. Unfortunately, $\chi(x)$ (hence $F(x)$) is far from being smooth; in fact, it is not even continuous and, in general, $F(x)$ does not belong to the class of $f(x)$. One may either smooth out $F(x)$ or use other procedures. Here the concept of isotropic discrepancy (see [20]) plays an important role, but this cannot be discussed here. The book by Hua and Wang handles these and related problems with greatest care. Perhaps the most remarkable feature of the book is the large amount of algebraic number theory presented and used. Of the ten chapters, more than half (Chaps. 1, 2, 3, most of 4, half of 5, all of 6) are devoted to basic number theory. The text starts with a thorough discussion of algebraic number fields and their units, followed by a study of certain symmetric functions. Next, the PV (Pisot-Vijayaraghavan) numbers are introduced (the algebraic integer α is a PV number if $\alpha > 1$, but for all its conjugates, $|\alpha^{(i)}| < 1$). The uniform distribution modulo one is carefully presented (but H. Weyl's name barely appears in a note at the end of Chap. 3). A relevant theorem of Vinogradov is proven and different types of discrepancies are defined and compared. Here the work of van der Corput, Hammersley, Halton, Hlawka, Zaremba, Niederreiter, and W. Schmidt is acknowledged, together with that of Korobov, the authors, Khintchine and Bahvalov.

Because of the great importance of the concept of the discrepancy of a set of points, we recall its definition. Let $P_n(k) = \left(x_1^{(n)}(k), ..., x_s^{(n)}(k)\right), 1 \leq k \leq n$ be a set of n points in the s-dimensional unit cube G_s, denote by $\vec{\gamma} = (\gamma_1, \gamma_2, ..., \gamma_s)$ any fixed point of G_s and set $|\vec{\gamma}| = \gamma_1\gamma_2...\gamma_s$. Let $N_n(\vec{\gamma})$ denote the number of points of $\{P_n(k), 1 \leq k \leq n\}$ that satisfy the inequalities $0 \leq x_i^{(n)}(k) < \gamma_i, (1 \leq i \leq s)$. Then

$$\sup_{\vec{\gamma} \in G_s} \left|\frac{1}{n}N_n(\vec{\gamma}) - |\vec{\gamma}|\right| = D(n)$$

is called the discrepancy of the set of points $\{P_n(k)\}$. If we have an increasing sequence of integers n_i, then we may compute $D(n_i)$ for each of these. If $\lim_{n_i \to \infty} D(n_i) = 0$, the sequence $\{P_{n_i}(k)\}(i = 1, 2, ...)$ is said to be uniformly distributed in G_s.

The text continues with the study of approximations by rationals and of the number of solutions of Diophantine equations and systems. This theoretical preparation is used (at the end of Chap. 4) to compute the discrepancies of several sets of points that are either g.p. or g.l.p. Here the author's own contributions play a prominent role (use of PV numbers and of generalized Fibonacci numbers, defined by $F_j = 0$, $j = 0, 1, ..., s-2$, $F_{s-1} = 1$, $F_n = \sum_{h=1}^{s} F_{n-h} n \geq s$).

Problems of uniform distribution are considered and functions of bounded variation in the sense of Hardy and Krause (already considered in this context by Zaremba [22]) are defined and studied. Let $\{P_n(k), 1 \leq k \leq n\}$ be any set of points with discrepancy $D(n)$. Then, if $f(\vec{x})$ is a function of bounded variation (henceforth always understood in the sense of Hardy and Krause) of total variation $V(f)$ where $f(\vec{x})$ is not necessarily periodic, it is shown that

$$\left| \int_{G_s} f(\vec{x}) d\vec{x} - \frac{1}{n} \sum_{k=1}^{n} f(P_n(k)) \right| \leq V(f) D(n).$$

The proof of this fundamental inequality is far from trivial and is given in full detail. There are two requirements that the points $P_n(k)$ have to satisfy in order to lead to useful quadrature formulae: (1) they have to have a low discrepancy (i.e., they should be distributed as regularly as possible); and (2) they should be easily computable. The authors list some 15 quadrature formulae, each with an upper bound of its error. Also, given any $P_n(k)$ in G_s, the authors construct a function $f(\vec{x}) \in C^\alpha$, ($\alpha = q + \lambda$, i.e., $f(x)$ has continuous derivatives up to order q and the last, no longer differentiable derivative satisfies a Lipschitz condition of order λ) for which

$$\left| \int_{G_s} f(\vec{x}) d\vec{x} - \frac{1}{n} \sum_{k=1}^{n} f(P_n(k)) \right| > c(q, \lambda, s) n^{-\alpha/s}.$$

This shows that, regardless of how well the points $P_n(k)$ had been selected, there is a limit on the precision of these quadrature formulae, beyond which we cannot hope to improve them, at least as long as we impose upon the functions $f(x)$ only smoothness, without periodicity. At this point, periodic functions are introduced and a norm $\|f^\alpha\|$ is defined. If

$$f(\vec{x}) = \sum_{\vec{m}} C(\vec{m}) e^{2\pi i (\vec{m}, \vec{x})} \text{ and } |C(\vec{m})| \leq C \|\vec{m}\|^{-\alpha},$$

then $f(\vec{x})$ is said to belong to the class $E_s^\alpha(C)$. Two other classes of functions, $Q_s^\alpha(C)$ and $H_s^\alpha(C)$, are also defined. They satisfy the inclusion relations $H_s^\alpha(C) \subset Q_s^\alpha(C) \subset E_s^\alpha(2^s C)$, but we shall not define them here more precisely. For functions of these classes, more precise quadrature formulae can be obtained than for nonperiodic functions. Hence, there is interest in reducing nonperiodic functions to periodic ones, and the previously mentioned three methods for this reduction are presented. Finally, the numerical integration of periodic functions is presented for each of the three classes defined and for selections of the $P_n(k)$ as p, g.p., or g.l.p. By estimating lower bounds for the error terms, it is shown that often the upper bounds of the errors are of the right order of magnitude, so that in general (i.e., for particularly bad functions of their class), the error cannot be further improved. This leaves open the possibility (in fact, the likelihood) that for the great majority of the functions to be integrated, the error is much smaller than the conservative theoretical upper bound obtained.

A full chapter (8th) presents numerical work. Specifically, g.p. and g.l.p. are computed by several methods, and the errors bounds to which they lead are compared to each other and to the actual errors in specific numerical examples. In this work, the generalized Fibonacci sequence is found particularly useful. Also estimates for the number of operations and computing time are made and are related to the obtainable accuracies. Frequent references are made to several papers published in Zaremba's collection [23], including at least one lengthy verbatim quotation. Several numerical tables in the appendix are explained and discussed. Also several conjectures are formulated to the effect that the results obtained are probably better than what we can actually prove by our present estimates of the worst-case error terms. Finally, as already mentioned, the last two chapters discuss interpolation for previously defined classes of functions and the numerical solution of differential and integral equations (both of Fredholm and Volterra type) respectively.

There is no doubt about the value and usefulness of this book—presumably the only available complete and systematic presentation of this important and interesting material. It is more to be regretted that the reading of the book is rendered difficult and at times actually unpleasant by certain shortcomings that may easily have been avoided. The book has no index. The translation is occasionally awkward (see, e.g., p. 27, lines 3–5). The proofreading does not appear to have been made with the necessary care; indeed, there are many printing errors and, while some are easily corrected by the reader, others may be quite confusing (e.g., the exponents r_i in Theorem 1.1 should be γ_i). Proper names are often misspelled (Minkowski, p. 33 last line; Hardy, p. 99, line 12; Korobov several times, occasionally (p. 86, line 9) as Kopobov, which suggests a translation from the Russian rather than from the Chinese). A section has the title "The Halton Theorem", but contains several theorems and Halton's is not identified among them. Some terms are used with a meaning different from their usual one (the sums of i th powers of the roots of a polynomial equation are called elementary symmetric functions). Often a symbol is not defined. Occasionally, it is easy to guess its meaning, like $\{x\}$ in Theorem 3.2 for the fractional part of x. Other times, like for (x) on p. 60, the reader may think first of the similar symbol defined in Theorem 1.5, where it means the group generated by x; this guess is wrong. The reader fortunate enough to know the book [14] by Kuipers and Niederreiter will realize that the symbol stands for the distance to the nearest integer. The reference to Theorem 7.4 (p. 155, line 6) should be to Theorem 7.14, etc. The examples could be multiplied. Some readers may wish to precede, or supplement, their reading of the Hua-Wang book by studying the papers by S. Haber, S. K. Zaremba, D. Maisonneuve, and H. Niederreiter in [23], all of which are quoted in the present book. A knowledge of the book [14] by Kuipers and Niederreiter, although this is restricted to the problem of uniform distribution, may also help in the reading of the present book. In spite of the mentioned superficial shortcomings, which should easily be taken care of in a new edition, the book by Hua and Wang is a most valuable contribution to numerical integration and to the solution of differential and integral equations. The book contains much material due to the authors themselves and, in many cases, the methods suggested have led to the most accurate results with a minimum of computations. The tables of the Appendix are valuable by themselves. Finally, the book itself is a brilliant illustration of the practical usefulness of pure, abstract number theory.

(See E. Grosswald, *L.K. Hua and Y. Wang, Applications of Number Theory to Numerical Analysis*, Springer-Verlag and Science Press, 1981. Bull. Amer. Math. Soc.; 1983, 489–496. References are omitted.)

The shortcomings mentioned in the review of this book are real. This is due to our long closure and lack of communication, and our lack of understanding of the terminology and notation used in modern algebraic number theory. Another reason is that the book was penned by me according to the ideas of Hua Loo-Keng, and was published too hastily without careful scrutiny, and there was no time to rework and republished it later.

6.7 Motivation

I have been very fortunate to enter the field of 'number-theoretic methods in numerical analysis' or 'pseudo-Monte Carlo methods'. In addition to paying close attention to the constructability and tractability of the algorithms we provide, it is also important to pay attention to the estimation of theoretical errors, which is a job of the same character as those we have done in analytic number theory in the past. It involved also many areas of number theory with which I was unfamiliar, such as algebraic number theory and Diophantine analysis, as well as knowledge of probability theory and differential equations. I read B.V. Gnegenko's *Tutorial on Probability Theory* and Wu Xinmou's *Tutorial on Partial Differential Equations* several times. This pushed me to study many areas of mathematics, learning as I went along.

We were fortunate to get a perfect two-dimensional integration formula (see (6.6)) when we first entered the field, and to have a way and framework for generalizing it to higher-dimensional cases that we could drill into over time to do it rigorously and completely. For me, this is indeed an extremely strong impetus.

I became interested in pseudo-Monte Carlo methods, and on the other hand, I never saw a direction in pure mathematics in which I could work so systematically. Therefore, pseudo-Monte Carlo methods became my only possible and best choice of research direction at that time. I never wavered from this direction, no matter what the political climate was: during the extreme left period or after reform and opening up.

Hua Loo-Keng wrote on his biographical outline that he was "pulled along by Wang Yuan" to show his approval of my proposal to engage in this direction. The original ideas and framework of cyclomatic field method and the PV number method belonged to Hua Loo-Keng and were developed by us together.

In 1965, Hua Loo-Keng's interest shifted to work in the direct service of the national economy. He wanted to popularize the "Program Evaluation and Review Technique" (PERT) and the "Optimum-Seeking Method" (OSM), the so-called "Dual Method", in the industrial sector of China. Our collaboration came to an end. Later, I continued the work and summarized it in a book, which was not completed until 1980.

As early as 1963, He Zuoxiu wrote a review of Hua Loo-Keng and Wang Yuan's booklet *Numerical Calculation of Integrals*, which was highly recommended. Feng Kang wrote a review of a revised version of the booklet and asked Wang Yuan to write an entry on Number-theoretic methods of numerical integration for the *Chinese Encyclopedia* (mathematics volume), and Wu Wenda also expressed his approval of our work. This is all encouragement and support for us.

In 1974, Hua Loo-Keng received an invitation to give a presentation on numerical integration at the International Congress of Mathematicians in Vancouver. He was very willing to go, and asked me to write a manuscript for him. I understood that he did not really care about giving a report, but could relax for a few days in the context of the Cultural Revolution. In fact, this was the third time for Hua Loo-Keng to be invited by the ICM. He was also invited to the ICM for his work on the theory of functions of multiple complex variables in 1956 and 1970. Hua, however, missed all three meetings because he did not get permission to participate.

Chapter 7
Lost Memories

Interview

The "Cultural Revolution" was launched in 1966. China since then had been haunted by 10 years of turbulence until the movement was stopped in 1976. For a whole decade, the country had been deprived of proper cultural and legal environment. Scientific organizations and colleges were closed down throughout the country and academic researches were deserted. What was your life like during this period?

7.1 Storm Coming

When the "Cultural Revolution" broke out, I was teaching at the University of Science and Technology of China (USTC). Hua Loo-Keng was appointed Dean of the Applied Mathematics Department (known as the Mathematics Department later) when USTC was established in 1958 and then Vice Chancellor of USTC. He offered the course of Advanced Mathematical Analysis for first-year students and asked me to be his assistant. For this reason, I worked as Hua's teaching assistant at USTC since its founding and helped him prepare handouts. But I felt awkward given that I was still employed at the Institute of Mathematics. As I was planning an official transfer, the Cultural Revolution started.

I was no stranger to political movements and had witnessed a handful of them, the Anti-Rightist Campaign, the Great Leap Forward, the Socialist Education Movement a.k.a. the "Four Cleanups Movement" and others, all of which seemed aggressive at the very beginning yet ended up nowhere, leaving the victims hurt and the persecutors discomfited while infusing doubt and misunderstandings among people. To tell the truth, I hated it. But I was scared and, consciously or unconsciously, tried to take a silent back seat, stay away from limelight, and keep silent. I chose to spend all my time with mathematics and wished to achieve something in this field. At first, this tactic worked and helped me avoid lots of troubles, for a time I was sheltered safe in this way. However, I had been at last subject to the impact of the decade-long political turmoil since 1966.

7.2 Regret and Sadness

The Institute of Mathematics took Hua Loo-Keng as the key target of criticism at the start of the chaos and I was thus dragged back by the "revolutionary masses" to receive criticism likewise.

Hua was asked to the Institute of Mathematics in August 1966 for the struggle session held at the gate of the south building of the Institute of Computing Technology. The movement leaders of the Institute of Mathematics arranged in advance all speeches to be delivered at the session. One of the leaders asked several students of Hua to give a joint speech and ordered Wan Zhexian, one of them, to write the speech and me to read it to the audience. At the meeting, the leader announced, "we'll allow Hua Loo-keng to have a seat so that he could take careful notes", and things didn't turn out too bad that time. I would, nevertheless, always feel sorry and guilty for reading the speech denouncing my teacher in public.

I chose to live a "carefree life" since then. After all, you'll never know tomorrow. Why not just take it one day at a time? Fortunately, I have been happily married with Ms. Guo Baowen since 1967 and welcomed our sons, Wang Xuan and Wang Ze, in 1968 and 1972, respectively.

On July 11, 1968, the Institute of Mathematics had a plenary meeting, calling for "Cleansing of Class Ranks", thereafter plunging itself into darkness. The "Capital Workers and People's Liberation Army Propaganda Team of Mao Zedong Thought" (the Team for short) was stationed in the institute in October 1968 and the cleansing campaign got even more violent. After some alleged "Capitalist Roaders" and colleagues with adverse political background had been dragged out for punishment, the cleansing, chiefly targeted at the young, had unexpectedly concentrated forces against the purported "Active Counter-Revolutionary Group" (the Group) which was said to be frequently convened to attack the "proletarian headquarter", in particular its key figures and their behaviors.

I was identified as a member in the Group after tough scrutiny and I had encountered the most tremendous hit on me during the decade. Everyone of the Group was tortured and interrogated in one way or another. A special investigation team was in charge of my case and asserted that I had not told the whole truth no matter how hard I tried to reflect on myself and sum it up in the self-statement materials. In retrospect, the one thing I'm proud of is, I have all along told the truth and nothing but the truth, there was no false accusation or backstabbing.

Finally, our "Group" was convicted after all the mess and the charge against me was "contradictions between people and the enemy to be handled as contradictions among people", which, after prolonged word-splitting, were further readjusted to "contradictions among people" at last.

At the end of 1968, the Team put its hand to cleansing the several chiefs of the rebel faction known as "the Torch Brigade", which was basically concluded in half a year. Some were later resettled to the CAS "Hubei Qianjiang May 7th Cadre School" for reform through labor, or "following the May 7th Road".

In the winter of 1969, I too was resettled there as the third batch of them.

7.3 Qianjiang Cadre School

The Qianjiang Cadre School, located at the Jianghan Plain of Hubei Province, was next to the worthless Jianghan Oilfield. It was a snail fever-stricken area and the staple crops were cotton and rice. In the school, there were several brick houses, each accommodating a dozen people and equipped with several wooden bunk beds and two desks. Different brigades had separate kitchens. I worked with the Fifth Brigade comprised of resettled cadres from the Institute of Physics and Institute of Botany.

Free from the torture, I was much happier in the cadre school. I lived with my old colleagues from the Institute of Mathematics who were well aware that the scrutiny over me made no sense and we got on well together. No one regarded me as an "Active Counter-Revolutionary". During the daytime we worked on the farm, mostly the cotton farm, and sometimes I went to do woodwork in the school workshop or carry bags on the drying yard. Besides I took on the task of building the cement road in front of the school. Nothing too hard for me.

We didn't eat well, of course. But it was much better than the two previous stays in the countryside. This time I had enough food everyday and a special treat from time to time. I remember sometimes we had shaomai, steamed pork with rice powder or dumplings. We improved our living conditions with our own hands. We had even dug a tap water well in the school so we could enjoy the tap water like urban residents. Also we had electric lights in the room.

In January 1970, during my first or second month in the cadre school, Lin Biao and Jiang Qing convened a meeting in the Great Hall of the People and gave the "latest instructions" on the "May 16th" issue. Some convicted chiefs of the Torch Brigade and other rebel factions were sent away for the scrutiny of people's dictatorship and cleansing of both their crimes and organizations. At nights they had to go to the struggle sessions held at the drying yard under the kerosene light. I was not responsible for the scrutiny nor the subject of the scrutiny and I was not even allowed to work for the movement. I had been there, done that and knew too well what it was all about. I had sympathy for those penalized "May 16th rebels", though I had ever been tormented by some of them. I hate political movements and it was ideal for me to stay away from the annoyance.

I had much spare time on my hands but not enough books to read. I learned hair-cutting at Zhejiang University and I took the time to offer haircuts for my team members, which brought us closer together. I learned playing Erhu in my high school so I was able to play for shows of our brigade and helped to make life more interesting.

During the school stay, everyone had to work in the countryside for a month, that was, to live with neighboring farmers and work with them. Actually, cadres like me in the school were sent to local families for arranged meals. Of course, we had to do some farm work to the best of our ability. The living standard was higher in southern rural area than that in the north, we could sleep in wooden beds, have rice, and live in brick houses. Life there was more difficult than that in the cadre school but we found it not too hard to carry on.

In the winter of 1970, we went back to Beijing for family reunion of the Spring Festival.

In August 1971, I was transferred back to the Institute of Mathematics following the official order.

7.4 Back to Normal

In 1971, the "September 13th Event" awakened people from momentary doubts and confusion and reminded us to reflect on the Cultural Revolution. Like dripping water piercing a stone, the event quietly prompted changes in people's attitude. My colleagues at the Institute of Mathematics came to their senses one after another and there were few real enthusiasts. All of us read from the newspapers like a recorder when having a meeting and copied word for word when making the big-character posters.

In fact, the Ping Pong Diplomacy cracked open the diplomatic door in April 1971 and C. Davis, the first American mathematician in over two decades was allowed in. He made a report on functional analysis at the hall of the Chinese People's Association for Friendship with Foreign Countries (CPAFFC) in a seminar hosted by Hua Loo-Keng. Then Chern Siing-Shen visited Beijing to meet old friends and gave several reports, among which the seminar titled The Essence and Significance of Mathematics was hosted by Hua Loo-Keng in Tsinghua University and attracted over 1,000 people.

Some mathematicians were invited to support the event, i.e., helping with reception and attending the seminars. Most of the May 16th counter-revolutionaries and the members of the Group from the Institute of Mathematics had been allowed to join, which indeed recovered reputation of themselves as well as everything concerning mathematics regarded as feudalism, capitalism, and revisionism during the "Cultural Revolution" and the "Anti-Rightist Movement".

It suddenly occurred to me that I had worked hard in and for the mainland since teenage years and achieved something in spite of the extreme hardships. How come I was identified as a counter-revolutionary at last, while the Chinese scholars living abroad for years were considered distinguished guests? I was at a loss. Moreover, our motherland spent a lot on offering a decent education and work environment to us in the hope of us paying back with our academic achievements, then how come the same motherland had launched one movement after another to distract us from work? Why? I just could not understand.

7.4 Back to Normal

The good teacher-student connection between Hua Loo-Keng and me was cut short due to the criticism against Hua Loo-Keng during the Cultural Revolution. In 1972, Hua Loo-Keng, back from Japan with a delegation led by Liao Chengzhi, gave me a call and said, "come to my home, I have something to share." Hua showed me *the Applications of Number Theory to Numerical Analysis*, collection of proceedings published in 1972 with S.K. Zaremba as the editor-in-chief, which included H. Halberstam's thesis on the Hua-Wang method. We share the joy over recognition for us two and the even bigger joy of reestablishing the trust and friendship of the old days.

The American Mathematician Delegation led by S. Mac Lane visited China from May 3 to 27, 1976, and spent quite a bit of time in the Institute of Mathematics. I delivered to the delegation a report entitled "Applications of Number Theory to Numerical Analysis", covering the joint academic research of Hua Loo-Keng and me.

At Hua Loo-Keng's invitation, the delegation went to see application scenarios of the "program evaluation and review technique and optimum-seeking method" (the "dual methods") in northeast China. Pollak invited him to write a book for the Birkhäuser on his experience of popularizing mathematical methods in China. After the Cultural Revolution, based on Hua's works and manuscripts as well as our results on "orebody geometry" and our experience popularizing linear programming in China in 1958, I wrote a book entitled *Popularizing Mathematical Methods in the People's Republic of China: Some Personal Experiences*, coauthored by us two and published by Birkhäuser in 1989. Unfortunately, Hua didn't see it. I translated the book into Chinese in 1991 and the new book, entitled *Selected Topics on Mathematical Modeling*, was published by Hunan Education Publishing House.

In 1978, Li Shangjie, the Party Secretary of the Division of Four Subjects, Institute of Mathematics, came to give back the materials I wrote during the Cultural Revolution. We didn't say a single word. I burned all the materials and thereupon wiped away memories of the past.

Chapter 8
Back to Math

Interview

During the "Cultural Revolution", the situation was better at the Chinese Academy of Sciences. In 1972, after Premier Zhou Enlai's intervention, who proposed to "strengthen basic science and theoretical research work," some academic research began to thaw, and individual scholars began to devote themselves to research in earnest. Can you tell us about the process and thinking of how to restart the work of mathematics research?

8.1 Selecting New Area of Research

In the summer of 1971, I was transferred back to the Institute of Mathematics from the May 7 Cadre School, and in September, the Lin Biao Event broke out, and since then, the research work at the Institute of Mathematics has been slowly resumed. The *Scientia Sinica* and *Acta Mathematica Sinica*, which had been ordered to stop publication during the "Cultural Revolution", were also gradually resumed.

I first spent half a month perusing the number theory section of the *American Mathematical Review* over the years, and browsing through the new foreign books on number theory, and learned that the achievements in the Diophantine analysis of and transcendental number theory were outstanding, especially the great achievements of W.M. Schmidt, A. Baker and S.A. Stepanov. In analytic number theory, K. Roth's improvement of the large sieve and E. Bombieri's proof of the mean-value formula for the distribution of prime numbers in an arithmetic series were outstanding works. My past work on analytic number theory has also been mentioned in a number of books and articles, which is quite gratifying to me.

On the other hand, I thought that after the completion of the theoretical proof of the so-called "Hua-Wang method", there was not much more to do in numerical integration that Hua Loo-Keng and I had collaborated on, and all that remained was to write some articles and books, so I was faced with the need to explore in another direction. This brings to an end of the collaborative research between Hua and me on number-theoretic methods in approximate analysis.

After reading many papers by Schmidt and Baker on the analysis of Diophantine analysis and transcendental number theory, I thought that some young number theorists should be taken on this path. I organized a small seminar, which included, in addition to me, Yu Kunrui, a student specializing in number theory at the University of Science and Technology, Zhu Yaochen and Wang Lianxiang, and Xu Guangshan, a student specializing in differential equations, who volunteered to work in the number theory group. I gave them the articles that I had read, and then they took turns to give presentations, just like in the Goldbach conjecture Seminar. They were thus better trained in this seminar. Later, they went to England, Australia, and Germany for further studies. They all chose the transcendental number theory as their study field.

After reading some of Baker's articles on transcendental number theory, I didn't have any ideas, much less see a topic I could work on in the long run, so I decided not to venture into this area of transcendental number theory.

In the summer of 1978, Schmidt visited Beijing, where he spoke about his recent work on the Diophantine equation and inequalities, which used the circle method and the estimation of the Weyl sum. These aspects happened to be familiar to me and thus drew my attention to his work.

In 1979, after a solo visit to France and West Germany, I moved to Dalham, England, to attend the International Conference on Analytic Number Theory. At the conference, I heard Schmidt speak again about his work, which deepened my impressions.

In 1980, I visited the United States as a member of the Delegation of Chinese Mathematicians. After the Delegation's work was finished, I visited the University of Colorado in Bordeaux for 2 weeks at Schmidt's invitation and stayed in Schmidt's house. I gradually developed the idea of combining Schmidt's method with Siegel's circular method for algebraic number fields, and thus extending Schmidt's results to algebraic number fields. This means that the idea of extending the results of analytic number theory to algebraic number field, which had been proposed by Hua Loo-Keng in the early 1950s, can now be tried to be realized.

In the early fifties, I read the works of Vinogradov and Hua Loo-Keng on the Waring problem. As a result of my study of pseudo-Monte Carlo methods, I have also worked on classical algebraic number theory and Diophantine analysis, which are my advantages. I could start to generalize Schmidt's results to the algebraic number field by reading carefully Siegel's article on the Waring problem in the algebraic number field and Schmidt's article on the minimal solution estimation of forms type. For me, the framework of this work is clear.

In the winter of 1980, I returned to Beijing from the United States. I immediately began to work in this area. I started by reading two seminal articles by Siegel on the Waring problem on algebraic number fields (1944, 1945) and two articles by T. Tatuzava on the Waring problem on algebraic number fields (1958, 1973). I learned

8.2 Extension of Schmidt's Results

about the circle method on algebraic number fields. I read two more articles by Schmidt, published in 1979, on minimal solution estimation for additive equations, and worked step by step.

8.2 Extension of Schmidt's Results

The so-called additive equation is

$$a_1 x_1^k + \ldots + a_s x_s^k = 0, \tag{8.1}$$

where $a_1, \ldots, a_s$ is a rational integer but not all of them are the same sign. By the circle method, Eq. (8.1) has a nontrivial solution in nonnegative rational integer $x_1, \ldots, x_s$, provided only that $s > c(k)$, where $c(k)$ denotes a positive constant that depends only on k and denotes a different value when it occurs in different places.

With respect to Eq. (8.1), we can also consider its minimal solution range, i.e., find an upper bound estimate for a minimal nontrivial solution that depends only on the coefficients $a_i (i = 1, \ldots s)$. In 1971, Pittman proved that if $s \geq c(k)$, then (8.1) has a nontrivial solution in nonnegative integer satisfying

$$\max_i x_i \ll \max(1, |a_1|, \ldots, |a_s|)^{c(k)},$$

where $c(k)$ are all calculable.

First for a special additive equation

$$a_1 x_1^k + \ldots + a_s x_s^k = b_1 y_1^k + \ldots + b_s y_s^k, \tag{8.2}$$

where coefficients a_i and $b_j (i, j = 1, \ldots, s)$ are positive integers, Schmidt proved that, given any $\varepsilon > 0$ and if $s \geq c(k, \varepsilon)$, then (8.2) has a nontrivial solution in nonnegative integers such that

$$\max_{i,j}(x_i, y_j) \leq \max_{i,j}(a_i, b_j)^{\frac{1}{k} + \varepsilon}. \tag{8.3}$$

The $\frac{1}{k}$ on the right side of (8.3) is the best possible. This is not difficult to learn from the following example: taking $a_1 = \ldots = a_s = a$ and $b_1 = \ldots = b_s = b$, where $(a, b) = 1$, so

$$\max_{i,j}(x_i, y_j) \leq \max(a, b)^{\frac{1}{k}}.$$

I understand that Schmidt used the circle method in his proof of (8.3), but the treatment of the "minor arc" is quite different from that of Waring's problem. Put simply, the circle method treats Waring's problem by directly using Weyl's sum estimate on the minor arc, whereas Schmidt proved that if

$$|L(\xi)| = \left| \sum_{P<x<P+T} e(x^k\xi) \right| \geq C,$$

Then there exist rational integers a, b, such that

$$||a\xi - b|| \ll \left(\frac{T}{C}\right)^{2^{k-1}} T^{-k+\varepsilon}, 0 < |a| < \left(\frac{T}{C}\right)^{2^{k-1}} T^{\varepsilon}, \tag{8.4}$$

Here, $\|x\|$ denotes the distance from x to its nearest integer.

Equation (8.4) indicates that there is a good rational approximation for ξ when Weyl's sum is large; otherwise, Weyl's sum is smaller (less than C).

Schmidt's article is written in a very clear and understandable way. By combination of his method with Siegel's circle method, I generalized (8.3) to an arbitrary algebraic number field.

Schmidt's second article proves that given $\varepsilon > 0$ and a set of rational integers a_i ($1 \leq i \leq s$) and suppose $s \geq c(k, \varepsilon)$, then the additive equation

$$c_1 a_1 x_1^k + \ldots + c_s a_s x_s^k = 0 \tag{8.5}$$

has a set of solutions, where $c_i = +1$ or -1 and x_i ($1 \leq i \leq s$) are totally nonnegative rational integers, not all zero, satisfying

$$\max_i x_i \ll \max(1, |a_1|, \ldots, |a_s|)^{\varepsilon}$$

I generalize this result to algebraic number fields.

Suppose $k = 2h$, if $c_i = +1$ ($1 \leq i \leq s$), then the Eq. (8.5) has no nontrivial solution in nonnegative integer when $a_i = 1$ ($1 \leq i \leq s$). Later I realized that the problem of solving Eq. (8.5), when $k = 2h$, is equivalent to solving the equation

$$a_1 x_1^k + \ldots + a_s x_s^k = 0 \tag{8.6}$$

in $x_i \in Z_+ \oplus e^{\pi i/k} Z_+$.

When I extended these two results to the algebraic number field, there was one point in the proof of my result where I could not get past. I am indebted to Feng Keqin for telling me about the Hurwitz Lemma so that the proof is complete.

In 1985–1986, I was invited by the Institute for Advanced Study in Princeton to visit for an academic year. I brought my father to live with me. I rented a unit in the Institute's dormitory, which had two bedrooms and one bath. My father and I each shared a room, and I cooked the meals every day.

This was a special year for the transcendental number theory and Diophantine approximation. Schmidt was invited to be a Professor of Excellence. I brought the manuscript on the additive equations mentioned above and asked him to review. He was kind enough to review it. I submitted both articles to the *Acta Arith.*, and both were published in 1987.

Schmidt suggested that I consider the problem of the Diophantine inequality. I read his article. He proved that,

Let $F_i(\vec{x}) = F_i(x_1, ..., x_s)(1 \leq i \leq h)$ be forms with real coefficients of odd degrees that are all $\leq k$. Given a positive number E, however large, there exists a constant $c = c(k, h, E)$ as follows. If $T \geq 1$ and $s \geq c$, then there exists a non-zero integer point $\vec{x} \in Z^s$, such that
$$|\vec{x}| \leq T,$$
and
$$|F_j(\vec{x})| << T^{-E}|F_j|, \ 1 \leq j \leq h,$$

where $|\vec{x}| = \max_i |x_i|$ and $|F|$ denotes the maximum absolute value of the coefficients of F, and the constants associated with $<<$ do not depend on T and F_j's.

If the coefficients of all forms F_j are rational integers, then it follows that given any positive number ε, there exists a constant $c = c(k, h, \varepsilon)$ as follows: If $s \geq c$, then there exists a non-zero integer $\vec{x} \in Z^s$ satisfying
$$|\vec{x}| << F^\varepsilon$$
and
$$F_j(\vec{x}) = 0, \ 1 \leq j \leq h,$$

where $F = \max(1, |F_1|, ..., |F_h|)$.

Schmidt suggested that I generalized the above result to algebraic number fields. He reminded me that for the totally complex algebraic number fields, the condition that the degree of forms be odd might be removed. His idea was right (it is easy to know for a single additive form equation). While at the Institute for Advanced Study in Princeton, I finished this research and submitted it as a paper to the *Journal of Number Theory*.

8.3 Book Review of *the Diophantine Equations and Inequalities in Algebraic Number Fields*

Upon my return to Beijing, I set about putting together a book on those researches and submitting it to Springer Verlag. It was reviewed by seven reviewers, all of whom considered it ready for publication. However, I still felt that it was not well written, so I put it off and rewrote it, and the book was published in 1991 by Springer Verlag.

In 1992, *the Bulletin of the London Mathematical Society* (BLMS) published a review of my book in its *Book Reviews* section as follows.

Suppose that I have a set of mathematical objects, which I wish to enumerate. I try to divide the set into two subsets in the following way. One subset can be enumerated (asymptotically) in a fairly straightforward manner. The other can be shown to have a smaller order using some averaging process.

This principle lies behind a very powerful method, known as the Circle Method, and it was first introduced into the additive theory of numbers around 1920. The Circle Method had its genesis with ideas of Hardy and Ramanujan in 1918 and was developed subsequently by Hardy and Littlewood. As an example of the results it produces, consider the equation

$$N = x_1^k + \cdots + x_s^k, \tag{8.7}$$

where $k \geq 2$, and N, s and the x_i are all nonnegative integers. Hilbert, in 1909, proved Waring's assertion that, for every positive integer k, there is an integer $s = s(k)$, such that every positive integer N can be represented in the form (8.7). Let $r_s(N)$ denote the number of such representations. The Circle Method yields an asymptotic formula for $r_s(N)$ of the kind (8.8) below. This is achieved by transforming $r_s(N)$ into an integral over the unit circle. The circle is divided ingeniously into two subsets according to the principle announced earlier. The following formula is obtained:

$$r_s(N) = \Gamma(1 + \frac{1}{k})^s \Gamma(\frac{s}{k})^{-1} \mathcal{B}(N) N^{\frac{s}{k}-1}(1 + o(1)), \quad N \to \infty \tag{8.8}$$

provided $s > f(k)$, for some reasonable function $f(k)$. In (8.8), the expression $o(1)$ denotes a function which tends to 0 as N tends to infinity. The expression $\mathcal{B}(N)$ denotes a function of N, s, and k, which has a positive lower bound independent of N. This is known as the Singular Series, and it has a property which is quite remarkable: the Singular Series satisfies an Euler product formula, and I shall comment on this fact at the end of the review.

The Circle Method has been extended and refined over the years, and in its various forms is applicable to many problems in additive number theory. The recent book by R. C. Vaughan (*The Hardy-Littlewood Method*, Cambridge University Press, 1980) is recommended as an interesting account of the progress made. It surprised me to discover that Siegel, as early as the 1920s, had attempted to generalize the Circle Method to deal with equations in algebraic integers. Only in 1945 did he succeed.

The book under review is a highly polished account of Siegel's method and its applications to Waring's problem and to other additive equations. Also treated are the results of the author on Diophantine inequalities for forms in many variables over algebraic number fields. These fundamental inequalities were obtained originally by W. Schmidt, using a variant of the Circle Method, for the rational field Q. The author remarks, rightly, that there is much to do in applying Siegel's method to generalize results to algebraic number fields, which have been obtained only for the rational field.

In the study of additive equations (such as Waring's problem) over algebraic number fields, there is a technical hitch, which needs to be understood. Consider

8.3 Book Review of *the Diophantine Equations and Inequalities ...*

the quadratic field $\mathbb{Q}\sqrt{d}$, where $d = 2$ or $3 \mod 4$. The algebraic integers in the field are of the form $a + b\sqrt{d}$, $a, b \in \mathbb{Z}$. The square of such an integer has an even second coefficient. Thus, an integer with an odd second coefficient can never be a sum of squares. It so happens that this is essentially the only barrier to studying Waring's problem for squares, and it is easily overcome in general. Given k, the ring of algebraic integers is replaced by the ring generated by the kth powers of the integers. This is the correct environment in which to study asymptotic formulae such as (8.8).

The first chapter of the book demonstrates the Circle Method as applied to Waring's problem over $\mathbb{Q}$. Apart from Vinogradov's refinement, the treatment is completely basic and very clear. This was a good decision by the author, and the chapter serves a dual purpose. First, the confidence of the reader is secured. Secondly, the strategy of the proof is laid down so clearly, that the task of introducing generalizations and refinements is made much easier.

The basic application of the method requires estimates for exponential sums and integrals due to Weyl and Hua. Over three chapters, these results are generalized very thoroughly to the context of an algebraic number field. The following two chapters set up the generalized Circle Method in the abstract. This method is then applied with increasing levels of difficulty to Waring's problem, to general additive equations then to the theory of small solutions of additive equations, and finally, in a thrilling finish, to the theory of diophantine inequalities for forms. In fact, the last two chapters represent the recent discoveries of the author, generalizing the results of W. Schmidt. The latter chapters contain results which require considerable technical skill. Thus, we have been guided, in less than 170 pages, from the basic principles to the most recent results in a technical field. That this has been achieved painlessly is due largely to careful planning and very clear exposition. The author is to be congratulated for his patient setting up of the general machinery. The book has a sensible pace, and I think that it is likely to become the standard text in applications of the generalized Circle Method. It looks good and is well priced at DM.98. For at least one of these facts, we are indebted to Dr P. Shiu, who processed the book using the TEX macro package from Springer.

We saw earlier that in applications of the Circle Method (for example, to Waring's problem), a condition of the kind $s > f(k)$ is inevitable. It is an important problem to determine how small $f(k)$ can be for the formula (8.8) still to be valid, in order that one may obtain an effective version of Hilbert's theorem. In the rational case, considerable progress has been made on this problem, even since the appearance of Vaughan's book. Apparently, the field is flourishing, and so there is certainly room for several advanced texts.

Now to return to my remark on the Singular Series. The gamma factors in (8.8) arise from an explicit computation of what is known in the trade as the Singular Integral. In all the treatments I have seen (admittedly, not many), the Singular Series and the Singular Integral are treated as quite separate objects. Compare this with algebraic and geometric number theory, where it is common practice to study complex functions, called L-functions, which have Euler product expansions. These are usually normalized by the multiplication of gamma factors. Originally, these normalizations were chosen because that is how "you get the functional equation to work." In 1950, in his PhD thesis, Tate altered the whole point of view of the subject by treating the gamma factors themselves as the "local factors at infinity." In this way, it transpired that the local arithmetic was basically the same at all the primes, finite and infinite.

Is it possible to apply Tate's ideas to the Circle Method? In other words, could the Singular Series and the Singular Integral be treated in one go by recognizing them as arising from local measures on suitable analytic spaces? In Chap. 7 of the book, Professor Wang calculates these objects in the general case, and both of them assume a very interesting shape.

(See Everest G.R. Book review: *Wang Yuan,the Diophantine. Equations and Inequalities in Algebraic Number Fields,*
Bulletin of the London Mathematical Society, 1992,24(4):83–94).

Chapter 9
Dabbling in Mathematical Statistics

Interview

You and Prof. Fang Kaitai won the second prize of the 2008 National Natural Science Award for "Theory, Methodology and Applications of Uniform Experimental Design", while the first prize of the same year was not awarded. In the introduction to the award-winning project, the Chinese Academy of Sciences states: "This project reveals the profound connection between classical factorial design, modern optimal design, supersaturated design, combinatorial design, and uniform design. The research work spans many different fields such as number theory, functional theory, optimization theory, experimental design, stochastic optimization, computational complexity, etc. It is an example of interdisciplinary intersection and penetration, and opens up a new research field. The project has published more than 80 papers and two monographs in English. The 40 papers and monographs listed have been cited 622 times in the Science Citation Index Database and 1512 times in the Chinese Science Citation Database, and the uniform design has been included in several international encyclopedias and statistical handbooks, and has been softwarized ..."

We would like to know how you and Prof. Fang Kaitai started your research collaboration. What do you think were the reasons for the success of this collaboration?

9.1 Origins

In 1975, I was asked by Fang Kaitai about a problem of numerical calculation of quintuple integrals arising in production practice. I introduced to him the number theoretic method of numerical integration. It worked well in use, and this gave him a good impression of the number theoretic method.

Fang Kaitai's original major was not mathematical statistics but operations research. He was very enthusiastic to go to the industrial sector to work on practical problems. Because he often encountered statistical problems, this led him to the path of mathematical statistics. And he had a very flexible grasp of statistics, i.e., he was able to always have intuition and ideas on how to deal with practical problems by statistical methods whenever he encountered them.

In 1978, Fang was working on a missile conductor problem, which simply means how to simulate the flight trajectory of a flying object based on the measured data so that the missile can hit the target. Because the calculation was too large, the existing computational and experimental methods were ineffective. This led him back to the number theoretic method, and he came to me for discussion.

With this problem as a background, we studied the following experimental design problem.

In an experiment, assume that there are s factors of which each has q (>1) levels. Ask how to structure the experiment.

If all possible experiments are done, there are a total of q^s combinations of factor levels. This is too many. The usual "orthogonal design" is to select rq^2 of these experiments to do, where r is a positive integer. However, when q sufficiently large, for example, $q > 10$, it is still too large.

We divide each side of the s-dimensional unit cube U_s into q equal parts, so that each test corresponds to a vertex of a small cube with side lengths $\frac{1}{q}$ in U_s. If "uniformity" is used as a criterion for selecting test points, it was known, from my previous study in number-theoretic methods for higher dimensional numerical integration, that a discrepancy of $O(q^{-1+\varepsilon})$ ($\varepsilon > 0$) could be achieved by selecting q single point. This means that it is sufficient to select $O(q)$ points out of the q^s total number of points.

First of all, in the past, the study of high-dimensional numerical integration of dimensions $s \geq 2$ required a relatively large number of points, and we have some such "large samples", but here we need "small samples", i.e., smaller q. Therefore, we need to find ways to transfer existing number theory methods to deal with the "small sample" problem. I decided to transfer the "good lattice method", whereby I worked with Fang to develop some "small sample" tables that could be used for arranging experiments. He also pointed out that the results of the experiment could be analyzed using regression analysis or stepwise regression. We named this experimental design method "uniform design" and used it for the above "conductor" problem, which was said to be very effective.

Next, we need to study the problem of formula experimental design (or mixture design). For example, if there are s factors $x_1, x_2, ..., x_s$ with $a_i < x_i < b_i$, where a_i, b_i are constants satisfying $0 < a_i < b_i < 1$, $1 \leq i \leq s$ and

$$x_1 + ... + x_s = 1 \tag{9.1}$$

Here, x_i can be the percentage of a drug in a chemical synthesis. Equation (9.1) means that the sum of the percentages is 100%, while $a_i < x_i < b_i$ means that x_i cannot be less than a_i and more than b_i.

9.1 Origins

This kind of question is common. I have been to Yanshan Petrochemical Company to investigate and I was asked this question by the technicians there.

Such problems were not dealt with in the book *Applications of Number Theory to Numerical Analysis* by Hua Loo-keng and me. All the studies there sought the sequence of points uniformly scattered on the s dimensional unit cube U_s. Several methods of periodization of functions were also presented there, equivalently to transform the domain of integration into U_s. Here the test region defined by the constraint (9.1) is a low-dimensional manifold ($s-1$ dimension) of s-dimension U_s. How to define the "uniformity" of the sequence of points on it and how to find the "uniformly distributed" sequence of points on it?

These two problems should have been considered in the *"Applications of Number Theory in Numerical Analysis"* by Hua Loo-keng and me, but I couldn't have thought of them if I hadn't dabbled in mathematical statistics.

Regarding the second question, fortunately, the domain of integration to be studied in statistics is concrete and does not need to be approached in an abstract way. For example, how to seek the uniformly distributed sequence of points on sphere

$$x_1^2 + \ldots + x_s^2 \leq 1$$

and spherical

$$x_1^2 + \ldots + x_s^2 = 1?$$

These all correspond to problems in statistics.

This kind of mathematical problem is not as hard as the purely mathematical problems I used to work on, and often after a period of effort, we solved the problem of how to define the uniformity of the sequence of points in these domains, and how to find uniformly distributed sequence of points. We have published several articles on this subject.

Immediately after the publication of these articles, they attracted attention in China. Engineers in some units redid the experiments done in the past by "orthogonal design" or other methods with "uniform design", and the results were not bad, while the number of experiments required was much less. In some industries, such as the pharmaceutical industry, they have spread uniform design and developed "application software". There are also many applications in the military sector, especially in the aerospace sector, but we don't know the results they get because of the secrecy surrounding their work. Fang pointed out that the advantage of the so-called orthogonal design is not really "orthogonality", but "uniformity". The uniform design is designed by capturing the key of uniformity, and it will naturally win the game.

9.2 Book Review of *Number-Theoretic Methods in Statistics*

Fang Kaitai and I decided to write a book to expound systematically the theory, methods, and applications of uniform design, which is entitled *Number-Theoretic Methods in Statistics*. In addition to publishing it in China, we also translated it into English and submitted it to three foreign publishers, all of which gave us successive replies agreeing to publish it. The British publisher Karpman Hall replied to us first and the English version of the book was published by them in 1994. Thus, the uniform design method is gradually gaining attention and application internationally. A book review is listed below.

It is a well-known fact that Monte Carlo methods are by nature asymptotically slow to converge. By spreading the abscissas "uniformly" in a deterministic way, these rates of convergence can improve quite significantly. This monograph presents ways to find a deterministic set of points uniformly scattered over an s-dimensional unit cube and shows how to use these sets instead of the random Monte Carlo method in a wide array of statistical problems. The applications presented include multidimensional integration, optimization problems such as nonlinear model fitting and maximum likelihood estimation, experimental design, computation of representative points for multivariate distributions, robust inference, tests of normality, and sphericity and projection pursuit.

I liked this book, even though I would have never anticipated its content from just reading its title. This is the first attempt to consolidate many years worth of research in a book format. Not surprisingly, the presentation leaves several loose ends that require further research. There are many ways of measuring the degree of uniformity of a given set of points. The authors review several of them in their first chapter and pick the Kolmogorov-Smirnov (K-S) statistic that quantifies the distance (they call it discrepancy) from the empirical distribution of the set of points to the uniform distribution. When dimension $s \geq 2$, it is very difficult to find a set with the smallest discrepancy. Number theory helps us find sets of points with asymptotically small discrepancies; hence the book's title. According to the K-S measure of uniformity, Monte Carlo sets are not uniformly scattered; my first surprise was to read that when $s \geq 2$, neither are the set of equilattice points; that is, the lattice with equally spaced abscissas in the unit cube of R^s. Yet according to other measures of uniformity, this grid of equally spaced points would be "legal." Why choose to minimize the K-S distance, taking a "minimax" approach, and why choose $O(n^{-\frac{1}{2}})$ as the minimum rate at which this distance has to decrease? The book does not really provide answers to these two questions.

This raises my main concern about the material. The authors use this asymptotic criterion that defines uniformly scattered sets only when computing bounds to the convergence rates for multiple integrations and for the search of maxima of functions. Thus, I fail to see the point in using those hard-to-compute sets when dealing with problems that require only a small number of points. Late in the book, (Sect. 4.7),

9.2 Book Review of *Number-Theoretic Methods in Statistics*

the authors scratch the surface of this question, and yet they stick to their number-theoretic-generated sets instead of mentioning the possibility of using simpler and more intuitive deterministic grids in some particular instances.

I am especially concerned about their naive approach to (model-free) experimental design problems in Chap. 5. They start framing it as if the universal goal when statistically designing an experiment was to choose the experimental conditions uniformly spread over the experimental region of interest—this is not the case. Then, in a very unfortunate Example 5.2, they try to push their approach through the following argument. The experiment has to consider the effect of six different variables on the response, and each variable must be considered at 17 (!) different levels that are not equally spaced. To look at all the 17^6 experimental conditions is out of the question. Even to run the 17^2 experiments required by a Latin square design could be too much. They claim that the way out would be to conduct a 17-run experiment on one of their number-theoretic-generated uniform grids in R^6. They neglect the fact that because the levels are not equispaced, their fancy grid is very far from being uniformly scattered in any reasonable metric, and that the sensible thing to do would be to forget about those 17 levels for each factor and start with a two- or three-level experiment. When it comes to small designs with many factors and perhaps more than two levels, I would rather stick with subsets of the equally spaced lattice of points (highly fractioned factorials and alike) than switch to their uniform designs.

When rates of convergence are an issue, however, such as in evaluating multiple integrals or in sequential optimization problems, I can see the relevance of considering these number-theoretic approaches as serious alternatives to Monte Carlo. When s = 1, the sets of lowest discrepancy are the sets of equispaced abscissas on the interval [0, 1]. That is, for the integration problems in R, their techniques are quite close to the classical formulas. For some of their examples in multiple integration, $(s \geq 2)$, I would have liked to see how the number-theoretic methods fared compared to the iteration of single integration done using Gaussian quadrature. I also felt that there should have been some reference to the work by P. Diaconis (1988) and O'Hagan (1992) on Bayesian numerical analysis. On the other hand, I was pleased to read how importance sampling and other variance reduction techniques work in their setting and to find hints to the possibility of mixing Monte Carlo methods with number-theoretic ones.

The book is self-contained, and it assumes a basic knowledge in calculus and statistics at the graduate introductory level. In addition to the main results, it is filled with much ancillary information that helps extend their techniques to many compact domains other than s-dimensional unit cubes and contains suggestions on how to use their approach in a sequential manner to even further enhance the convergence rates of these optimization tools. The book comes filled with many multivariate distributional results and results useful when dealing with particular transformations of random variables. Each chapter has exercises that complement the main theoretical results, but real sets of data are nowhere to be found in the text.

It is unfortunate that this text is plagued with plots that do not have any labels on an axis, symbols meaning two completely unrelated things, incomplete tables, sentences that lack verbs, typographical errors, syntactic mistakes, and calls to the wrong theorems and equations. Addicts to scientific rigor will suffer repeated handwaving statements, missing regularity conditions, and the presentation of statistical models without the specification of their error structure. In one instance (p. 124), they even fall into the trap of implying that $E(\log y) = \log E(y)$. And yet, despite these shortcomings, I found this to be a highly thought-provoking book. When reading about the nontrivial ways that these techniques generate deterministic uniformly scattered grids, the reader should always bear in mind that the generation of uniform random numbers used in Monte Carlo is not a trivial matter. The main difference is that random number generation routines have been in our toolboxes for quite a while now. Generators of uniformly scattered deterministic sets of points will have to be made available in scientific libraries before these number-theoretic methods become a real alternative in our daily work. Despite the drawbacks mentioned above, I was pleased to be introduced to this new material and feel that Number-Theoretic Methods in Statistics is a valuable addition to the statistical literature. (See Josep Ginebra, "K.T. Fang and Y. Wang, *Number-Theoretic Methods in Statistics*, Chapman and Hall, 1994"; J. of Amer. Sta. Asso.; Sept., 1995. 1134.)

The shortcomings of the book are as the reviewers have pointed out. We have not worked carefully enough and we have published it too hastily.

9.3 Implications

As the range of applications of uniform design expands, it can be used for optimization problems, nonlinear model fitting, maximum likelihood estimation, robust inference, and projection tracing, in addition to computing high-dimensional integrals and experimental designs. It is often reported in the press, which has attracted the attention of leaders.

After reading the material in *Uniform Design* and *Selected Papers on Applications of Uniform Design*, Professor Tsien Hsue-Shen, the most famous aerodynamicist and aeronautical engineer in China, had written to Professor Zhu Guangya, the famous nuclear physicist and then as Chairman of the China Association for Science and Technology, as follows.

Chairman Zhu.

You must already have this copy, I have one extra copy and think it is important and should be promoted, so I offer it. The China Uniform Design Association should also be supported.

Please use your discretion.

Tsien Hsue-Shen (November 8, 1993)

Tsien wrote to me again on November 30, 1993, to offer encouragement and support.

Comrade Wang Yuan

I have read many times in the press recently about the "uniform design" method created by you and researcher Fang Kaitai, and I am very pleased to learn of its great

9.3 Implications

significance in practical application! I am writing to extend my heartfelt congratulations to you and Comrade Fang Kaitai. I congratulate you on your important contribution to the country and to the progress of mankind in the world.

How is your health? I wish you a speedy recovery!

With best regards!

Tsien Hsue-Shen

In September of 1994, Professor Zhou Guangzhao, then President of the Chinese Academy of Sciesnce, made an inscription for the uniform design method.

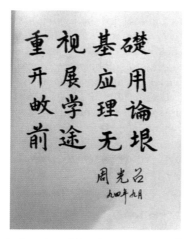

Zhou Guangzhao's inscription Attaching importance to basic research And developing application, Mathematical theory is of great expectation (left)

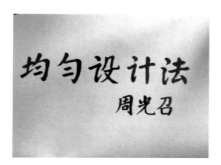

The uniform design method (right)

Zhu Guangya specifically invited me to give a presentation at the Science Hall, and he personally presided over it. I was greatly encouraged.

In 1994, with approval, the Uniform Design Branch of the Chinese Mathematical Society was established, affiliated with the Third Academy of the Ministry of

Space, with Fang Kaitai as chairman, Zhang Jianzhou as secretary-general, and me as advisor. We were invited by several research units of the aerospace and navy departments in Beijing, to give presentations and also held short workshops. By this time, Fang Kaitai had left for Hong Kong to work at the Baptist University, so Zhang Jianzhou and Liu Xiufang, a professor at Beijing Normal University, often went with me to give lectures. First, I gave a 1-hour presentation on the overview and academic framework of uniform design, while Zhang and Liu talked about details and applications. On many occasions, I was accompanied by Han Feng, Deputy Director of the Institute of Mathematics then.

In 1998, Shing-Tung Yau initiated and chaired the International Congress of Chinese Mathematicians and founded the journal Mathematics and Humanities. He invited Fang Kaitai to give a presentation at the second conference and to write an article on the uniform distribution. He gave importance and encouragement to our work. There are also successful applications abroad, for example, Ford Motor Company in the United States used uniform design in the initial development of cars and the University of Maryland used it in cancer cocktail drug trials. Fang Kaitai was invited to visit Ford, and the United States also programmed the uniform design method into statistical packages. The Handbook of Statistics and the Encyclopedia of Statistics, which have been published abroad, describe uniform design in detail and have asked Fang Kaitai to write a dictionary entry. This was gratifying.

9.4 Gentlemen's Agreement

Fang Kaitai and I have had a very successful collaboration. Our collaboration was not a joint effort to tackle a difficult problem, but a complementary relationship. I do mathematical problems and he does statistical problems. The reader can tell at a glance which work was done by whom. Therefore, the five articles that are mainly mathematical are signed in the order of Wang and Fang, and the books and articles that are mainly statistical are signed by Fang and Wang. Although I have repeatedly said that some of the works, to which I have no contribution, need not be signed, Fang insisted that I sign or let me write the foreword to his book. This shows his high moral character. Our cooperation was always pleasant.

Fang Kaitai's approach to learning statistics is to learn by going to the factory site to solve practical problems. This was very inspiring to me, while my way is to grasp the problems and cases he raised to think and learn, so that I can also give presentations on the theory and application of uniform design in Hong Kong, Taiwan, and abroad. In fact, I have not even read a statistical book.

Although statistics require mainly "small samples", I think it is sufficient to work out a "good" sample by approximation. For "very small samples", the computer can calculate the sample of the best discrepancy, but there is not much improvement. Also, the definition of disperation could be modified. I won't get involved in these mathematical issues.

9.5 Coincidence of the End and the Beginning

When Fang Kaitai and I started working on the uniform design, his question about the estimated area of destruction of the bomb always kept me hanging on to it. In mathematical terms, it can be described as follows.

Suppose there is a fixed circle O, and other m random circles of equal area $O_1, ..., O_m$. Ask,

what is the distribution of the area S so that $(O_1 \cup ... \cup O_m) \cap O = S$ (here, we will still denote the area of S by S)

For $m = 1$, we can express the true value of S by the length of the line connecting the center of the circle O with O_1. When $m \geq 2$, it becomes difficult to find an analytic expression for S. For this reason we had held a seminar at the Institute of Applied Mathematics devoted to this problem. For $m = 1$, Fang Kaitai performed computer simulations using both the set of equidistant lattice points (classical net) and the set of good lattice points (number-theoretic net) and proved that the good lattice points are much better.

Of course. applied mathematics is allowed to use simulation. That is, to use some examples to compare different methods. But wouldn't it be better to illustrate it with "theory"?

In 2008, it occurred to me why not make a mathematical statement using results from discrepancy estimation and geometric number theory.

Taking the Gauss problem of integer points in a circle as an example, the discrepancy of the classical net cannot be better than $O(x^{-\frac{3}{4}} \log^{\frac{1}{4}} x)$. While the number-theoretic net discrepancy (Fibonacci sequence) can reach $O(x^{-1} \log x)$, etc.

I have written a short article. It is a purely number theory article. Without statistics, however, I would not have thought of net other than classical net, and it was deserved to work with Fang Kaitai.

In 2016, I suddenly received a number of emails from foreign magazines asking me to publish more details, and I forwarded one of them to Fang Kaitai. I soon got hospitalized with pneumonia and haven't read emails since. In fact, these results are obvious once one is familiar with geometric number theory and pseudo-Monte Carlo methods.

It's a coincidence that I engage in uniform design, starting and ending with the same problem.

Chapter 10
A Sip of Mathematics History

Interview

You edited and published *Goldbach Conjecture* (English version) in 1984 and the second edition in 2002, which has been recognized as the best book on this subject. You compiled and published *Hua Loo-Keng: A Biography* in 1995 and the revised edition in 1999. Its English edition was published in the same year by the Springer. It's a seminal book well-recognized by academia and people home and abroad. As a mathematician and maths historian, would you talk about how you got the idea to research on maths history and what you've been through in editing and writing the two books above? Meanwhile, we noticed that you have no more books in this field after *Hua Loo-Keng's Mathematics Career* published in 2000 by you and Professor Yang Dezhuang. Are you working on another masterpiece, or do you just quit from research on maths history?

10.1 Goldbach Conjecture

A mathematician, whatever his/her research interest may be, should have some understanding of history of mathematics, in particular the history of his/her own area. Maths history should therefore be the compulsory course of all mathematicians. But to be a historian on maths is another thing.

As early as college years, I was intensely interested in maths history, but I never thought I would take up research in this field. As early as the 1980s, I had the thought for the first time that there should be an objective description and evaluation of the history of Goldbach Conjecture and in particular the introduction and development of thoughts and methods brought up for the research, to which important documents should be attached.

The thought may have started earlier. Once I suggested Pan Chengdong write a book on Goldbach Conjecture for the Springer. He and his brother Pan Chengbiao spent quite a bit of time and completed a great book. But the publisher thought it failed to meet the standard because the Three Primes Theorem and Chen's theorem, the main achievements on Goldbach Conjecture, had been published in many other books. Of course, they were biased. In fact, the Three Primes Theorem had been cited as applications of circle method and those of the exponential sum estimate of prime variables, while Chen's theorem the applications of the sieve method. There was no such a comprehensive monograph based on Goldbach Conjecture and its variations, instead of as applications of certain methods. I think such a book deserves to be published. It was in fact what Hua Loo-Keng had in mind when he asked us to write a monograph on Goldbach Conjecture for series Type B.

I decided to change my way and write a maths history book on the subject.

To begin with, I wrote a detailed introduction, stating the developments of main methods in studying Goldbach Conjecture, in particular original ideas in this field. Over 20 important papers or excerpts were included in the book, titled *Goldbach Conjecture* and edited by myself. It was published by the World Scientific in 1984 and well received afterward. For example, a mathematics dictionary published in the former Soviet Union referred to this book for its entry "Goldbach Conjecture". In 2002, at the invitation of publishing house, I made some supplements and republished the book, that was the second edition.

10.2 Hua Loo-Keng's Outline

In the early 1980s, H. Götze, the head of Springer approached Hua Loo-Keng with the notion of publishing his book. Hua replied he was aged and asked Götze to discuss it with Wang Yuan. Götze then talked with me, saying he would publish Hua's book *The Estimation of Exponential Sums and Its Applications in Number Theory* after supplementing as the volume *Analytical Number Theory* of Springer's *Encyclopedia in Mathematics*. The book was co-authored by Hua Loo-Keng and me. I declined in view of the workload. Götze also proposed to publish *Loo-Keng Hua Selected Papers*. I was all for it of course. Hua planned to put the collection of papers in four parts, i.e., number theory, algebra and geometry, complex analysis, and miscellaneous. There needed to be an introduction for each of the first three parts. Hua asked Wan Zhexian, Gong Sheng, and Lu Qikeng, and me to write one respectively. After reading all of the three introductions, he told me, "They're OK, but my favorite is yours." I responded immediately, "I could write a biography for you in the future."

In spring 1985, Hua asked me to go to his house at Chongwenmen. It took over an hour to get to Chongwenmen from Zhongguancun by bus. When I arrived, I saw him lying in bed, looking fragile. He said, "I knew it would take more than an hour, so I draft an outline of biography for your reference, what do you say?" The outline was written in the margin of a piece of scratch paper:

10.3 Hua Loo-Keng: A Biography   97

(1) Complete trigonometric sums, Davenport forgot he was the referee!
(2) Esterman's doubt, changed a few words, source for Linnik and Hilbert in Davenport's words.
(3) Tarry problem.
(4) Influence of Vinogradov.
(5) Rising influence of contemporary mathematics in China.
(6) Blocking and change of direction, matrix geometry, and automorphic functions.
(7) The fundamental theorem on semi-automorphism and projective geometry.
(8) Cartan-Brauer-Hua theorem.
(9) A problem solved within a meal time.
(10) After 1950.

>   Introduction to Number Theory
>   Classical group
>   Several complex variables
>   Introduction to Advanced Mathematics
>   Starting with unit circle
>   (10') being dragged to a road by Wang Yuan

(11) Caring for youth.
(12) Cultural Revolution, home searched, manuscript lost, program evaluation and review technique, optimum-seeking method, go across China, engaged in applied mathematics while thinking about pure mathematics.
(13) International maths conference in August 1980.
(14) Through thick and thin, fair and foul.
(15) Mathematical Seminar at Chicago.
(16) Old and new friends.

Item (13) of the outline refers to the International Symposium on Analytic Number Theory held in Durham, UK, July 22–August 1, 1979. There were two (14)s and two (10)s, but the second (10) was marked as (10').

From the outline, all to be written in the biography were the maths researches he listed, besides persecution suffered during the Cultural Revolution, popularizing of the "dual method", CPC membership and guidance for youth. One thing he emphasized was not to mention his childhood. The biography should be based on his maths life.

10.3 Hua Loo-Keng: A Biography

In 1985, Hua Loo-Keng passed away in Tokyo, Japan. I got down to writing *Hua Loo-Keng: A Biography* right away. I decided the book, besides the contents he listed, should be an overall reflection of Hua's legendary life, including the hardships and misfortunes, research experience, academic achievement, personal charisma, and patriotism. The book, themed Hua Loo-Keng, should partly reveal the developments and background of modern mathematics in China and contribute to relevant researches.

The first thing was to gather information extensively. I learnt about his life experience in *Jintan Historical Accounts* Volumes III, VI, and VIII.

During a visit to the Chinese University of Hong Kong, from *Biographical Literature* published in Taiwan and some Chinese newspapers and magazines published overseas I learned his work experience at Tsinghua University. Also I looked through newspapers published in the mainland for some of Hua's speeches, especially the collection of big-character posters during the Cultural Revolution by experts overseas. I had serious consideration and verification on the materials to make sure they were true and reliable.

Hua studied and contributed extensively, but I knew merely part of it. Of course, I knew his work on number theory and its applications, but little was known about his work on complex analysis, algebra, and geometry. Fortunately, the *Loo-Keng Hua Selected Papers* had been published. In addition, I had *Classical Group* and *Harmonic Analysis of Functions of Several Complex Variables in the Classical Domains* to read for reference. It took me a year to look through those materials. Xu Yichao helped me many times during the preparation, which led me to an overview of Hua's researches on complex analysis, algebra, and geometry as well as the background and interrelations of those researches.

I interviewed Lu Ruqian for the conditions in the Institute of Mathematics during the Cultural Revolution. For Hua's popularizing of the program evaluation and review technique and Optimum-Seeking Method, I interviewed Ji Lei. Both Ji and Lu had memories great enough to remember details and even specific dates of certain events.

And after Hua's death, I paid special attention to eulogies and memorial essays published home and abroad.

When I had a wealth of materials in hand, I got started. When one or two chapters were finished, I would put the draft aside and stop the writing. After a year or so, I would get back to the draft, read for a couple of time, and made supplement, revisions, and embellishment before I continue to write one or two chapters more.

In this way, I had the complete draft several years later. After reading through the draft for several times, I found the original length of 600,000 or so characters too large and therefore cut the draft radically or directly deleted and rewrote some of the chapters. I finished the draft 7–8 years later. I simply invited several friends to have a read.

I decided there would be neither debut party nor preface and endorsement by celebrities, the book would be put to market for the readers to comment.

In 1994, I offered the draft to Kaiming Press and Taiwan Chiu Chang Mathematics Publisher, which published the book in simplified Chinese and traditional Chinese, respectively.

It had won considerable popularity among mathematicians in mainland, Hong Kong and Taiwan. The CCTV program Reading once gave a half-hour exclusive interview for this book, for which the hostess asked questions and I answered. The interview had been broadcast for three times during the sessions of the National

People's Congress and the National Political Consultative Conference, lots of NPC deputies and CPPCC members watched it on TV.

Shortly afterward, the Springer Singapore decided to publish the English version of *Hua Loo-Keng: A Biography*. They invited British mathematician P. Shiu as the translator, who had migrated to Britain from Hong Kong since childhood. He didn't speak Chinese but could read Chinese books and newspapers. His wife was British and a literary scholar, who helped him with the translation. Shiu did a great job. The English version was published in 1999. The Springer Tokyo decided to publish the Japanese version and invited mathematician Shingo Murakami as the translator. Springer Tokyo and I signed a contract for the publishing. The translation had been finished soon but the publishing was long overdue. I didn't bother about the issue any longer, probably because I thought the English version was good enough. There were some more Japanese publishers planning to publish the Japanese version afterward. They got in touch with me through Qu Anjing and I therefore entrusted Qu with it. The Japanese version was not published at last, maybe because there was the English version already.

Kaiming Press printed 15,300 copies in total. With my permission, Hope Stacks had printed 10,000 copies free of charge. After that, I transferred copyright of the book to Jiangxi Education Publishing House and made further supplement and revision. They had a print run of 5,000 copies. Chiu Chang Mathematics Publisher of Taiwan printed 4,000 copies in two versions. Baihua Literature and Art Publishing House of Tianjin signed a contract with me to republish the book, which was out of stock long ago. We failed to fulfill the contract due to the disapproval of the Tianjin Publishing Bureau. The good news was that Dalian University of Technology Press decided to print over 1,000 copies of *Hua Loo-Keng: A Biography* to mark the centennial anniversary of his birth.

The book had won popularity and encouragement among experts and people, many of whom, including 100 or so professors, wrote to convey congratulations and appreciation. Just to mention a few, please see the letters of Su Bu-Chin, Tsien Hsue-Shen, Chern Siing-Shen, and Yang Chen-Ning as well as the e-mail of Shing-Tung Yau as follows:

Dear Academician Wang Yuan my comrade

Wishing you all the joy of the season and a new year fully loaded with happiness.

I've received your book *Hua Loo-Keng: A Biography*. Thank you for remembering, the book is a special gift to renew our warm friendship.

Mr. Hua has passed away, but his accomplishments stay forever. I learnt a lot from your book, as you can see, a man at his 90 s could still learn new tricks.

Best regards and good luck with your studies!

Su Bu-Chin
Jan. 19th, 1995

Dear Academician Wang Yuan,

Your letter on Jan. 10th received. So was your book *Hua Loo-Keng: A Biography* in care of Comrade Qin Wenliang. Thank you for that.

I'd like to extend my congratulations on your winning the Ho Leung Ho Lee Award and my gratitude for the regards you asked Comrade Wang Shouyun to convey during the award ceremony in Great Hall of the People on Jan. 12th.

I'll read your book carefully.

Thank you for your offer to write my biography. In my opinion, I already have a biography of over 10,000 words, written by Wang Shouyun and other comrades. That's enough.

Best regards!

Tsien Hsue-Shen
Jan. 17th, 1995

To Wang Yuan,

Thanks for your book. Hua had a brilliant life and now he is a part of history. In retrospect, I have mixed feelings. The book would be an important part of China's science history.

I'll arrive in Beijing on April 21st. Can't wait to see you soon.

Regards

Shiing-Shen
Mar. 15th, 1995

To Wang Yuan my dear brother,

Pleased to have your letter. *Hua Loo-Keng* is a great book. I learned a lot.

Hope to see you in the USA, Hong Kong or Beijing soon.

Have a nice summer.

Chen-Ning
July 15th, 1991

Dear Professor Wang,

I enjoy your book on Hua.

Shing-Tung Yau
Aug. 6th, 1999

Qian Wenzao told me the book received praise from Tsien Hsue-Shen

10.4 Hua Loo-Keng's Mathematics Career

Jiangsu TV had produced an eight-episode series based on the book *Hua Loo-Keng: A Biography*.

The book won gold award in the first Wu Dayou Popular Science Book Award of Taiwan. I went to Taipei to accept the award, presented by Yang Chen-Ning.

I referred to all materials and statements available during the book writing. Those materials didn't come easy and therefore we should make the best of them. I discussed with Yang Dezhuang, a graduate from the University of Science and Technology of China, that maybe we should edit and publish those materials. Yang was familiar with Hua's work on popularizing mathematics and could offer memoirs and comments in this regard. Finally, we decided to jointly compile a book titled *Hua Loo-Keng's Mathematics Career*, which was published by the Science Press in 2000.

10.5 Done and Dusted

The question in front of me was, shall I carry on with my research on history of mathematics? First of all, I had an analysis of my advantages in this respect—I'm a part of the research area and a part of China's modern mathematics. I know the Goldbach Conjecture well and I've already given a full play to this, while other areas I've studied appear too small to be written into history books. Moreover, in comparison to Hua Loo-Keng, I could not find another mathematician of the same importance in terms of academic achievements. I think I've run out of my advantages, that is to say, I'm done with the modern math history.

During my writing of *Hua Loo-Keng*, I got familiar with part of the mathematical achievements in modern China. Maybe I might consider writing on China's modern mathematics history. But it may take 6–7 years to finish the book and I'll have to learn lots of unfamiliar maths knowledge and interview lots of experts and insiders, more than that, it involves evaluation and weighing of mathematicians in different areas and their work. It's too big a bullet to bite. Besides, the writing, subject to personal preferences, could be biased and unfair and may incur dissatisfaction. I flinched and resolved to quit my research on history of mathematics.

Chapter 11
When the Curtain Falls

Interview

Generally speaking, a collection of works or selected papers of a scientist would take in the masterpieces and major pieces during the scientific career, which defines a scientist's work through his/her lifetime. In 1999, Hunan Education Publishing House published *Selected Papers of Wang Yuan*, taking in your papers during the decades-long span from 1956 to 1998, which is a fairly complete reflection of the developments of number theory in China. In 2005, the World Scientific Publishing Company published its English version *Selected Papers of Wang Yuan*. Would you please talk on how both were published and the feedback of the academics?

Besides, you've done a lot in scientific popularizing. You've been translator, author, and editor for a lot of books and developed an interest in calligraphy in later years. Would you please elaborate on it?

11.1 Selected Papers of Wang Yuan

Records of my study on mathematics are to be found in Chaps. 5, 6, and 8, 9 and 10 of this book. One thing in common for all the areas I've studied is I started early, when the area was still at its early stage of development and it was easier to be productive. The later, the more difficult it gets.

Maybe it's time to call an end to my study on maths. One thing left undone was to publish collection of my works. As for my research on maths, is there anything worth it? I'll just leave it to coming generations. From now on, I'd do whatever I love in my power.

In the late 1990s, Meng Shihua from the Hunan Education Publishing House (HEPH) told me they were to publish *Selected Papers of Wang Yuan*. For a mathematician, it's extremely important to get works published, and even more important to get collection of works published. I felt honored on the one hand and anxious on the other. How much exactly is my work worth? I had been reflecting on and reevaluating my work since HEPH brought it up. I found it not original enough. However, I gave my consent to the publishing at last. Undoubtedly, it would be a mirror for my work and I was eager to see the reflection. I invited Yang Lo and Pan Chengdong as the editors. Yang wrote a preface for the book, which was published in Chinese in 1999 and remained in obscurity since then.

It happened that the Guangdong Science and Technology Press (GSTP) and Springer Singapore (SS) as business partners were planning to jointly publish collection of works in English for Chinese mathematicians. Zheng Lihua, editor working with the GSTP came to ask for my opinion. I replied, "I'm in favor of it. But please make sure that Springer Singapore decide the namelist, you shall not make the decision."

Shirley, then head of SS, wrote to me later, telling me they were going to work with GSTP and publish collection of maths papers of Chen Jingrun, Lu Qikeng, and me in English. They planned to launch the collections of us three simultaneously to build momentum and asked me to help contact Chen Jingrun and Lu Qikeng.

Things got complicated. *Selected Papers of Chen Jingrun* (Chinese Version) had been published by Jiangxi Education Publishing House(JEPH), while mine was published by HEPH. Would the above two publishers transfer the copyright for English version to SS and GSTP? Chen Jinrun had passed away, so I told Lu Qikeng about this. At the same time, I wrote to ask JEPH and HEPH for their opinions. They declined. The trio publishing plan was thus made impossible.

Therefore, I suggested SS and GSTP jointly publish Lu Qikeng's collection (English version), while JEPH and HEPH work with SS to publish collections (English version) of Chen Jingrun and mine, respectively. SS did not agree with my plan and insisted it would be a trio. We didn't make it at last due to disagreement.

Then Shirley sent me a contract, specifying that SS would have an independent contract with me and publish my collection in English. Still, I was worried that my works were not original enough, in addition, the copyright was still in the hands of HEPH. I thought it better to make decision after a while, so I did not give my consent. However, I took the time to translate the articles in *Selected Papers of Wang Yuan* published by HEPH into English. Shortly, the SS was incorporated into Springer Hong Kong and Shirley changed her job. Springer Hong Kong had no intention of publishing the English version and the plan was thus dropped.

In 2004, Xu Zhongqin called to tell me that the Singapore-based World Scientific Publishing Company (WSPC) planned to publish the English version. I gave my consent right away as long as WSPC sent me an official invitation and obtained the copyright from HEPH.

I wrote to HEPH editor Meng Shihua and Meng promised to transfer the copyright for free. Meanwhile, I received the formal invitation signed by Phua Kok Khoo, WSPC editor-in-chief. Since I already had the translation (basically in accordance to the Chinese version except for a few articles) in place, the draft would be ready for submission with a little proofreading. Considering the flattering words on my career in the preface of the Chinese version, which I understood was the editor's duty, I decided to be the editor myself for the English version and wrote a preface as follows:

Earlier in 1998, Professors Pan Chengdong and Yang Lo, and many of my friends and colleagues urged and encouraged me to publish a volume of my selected papers since most of these papers were published in China and hard to find elsewhere. The Hunan Education Publishing House and in particular, Ms. Meng published a Chinese edition which appeared in 1999.

Now World Scientific Publishing Company and Dr. Phua have invited me to publish an English version of my selected papers. This is really my honor and I accepted the offer. Ms. Tan and Ms. Chionh helped me with the editing work.

I have been working in Chinese Academy of Sciences since 1952 when I graduated from the Department of Mathematics, Zhejiang University. My teacher, Professor Hua Loo-Keng, led me to the field of Number Theory. We also cooperated for a long time on the applications of number theory to numerical analysis. My other longtime collaborator is Professor Fang Kaitai. Our joint work is in experimental designs. In order for the readers to understand my life and works, the volume contains a related paper of Professors Li Wenlin and Yuan Xiangdong.

The readers can also see that my works are related and influenced by N. C. Ankeny, N. S. Bahvalov, A. Baker, V. Brun, A. A. Buchstab, D. A. Burgess, J. R. Chen, T. Cochrane, H. Davenport, P. Erdös, Fang Kaitai, G. H. Hardy, Hua Loo-Keng, N. M. Korobov, P. Kuhn, Yu. V. Linnik, J. E. Littlewood, T. Mitsui, Pan Chengdong, W. M. Schmidt, A. Selberg, C. L. Siegel, and H. Weyl.

Finally, I would like to take this opportunity to express my sincere thanks to those colleagues and institutions for their kind help.

Wang Yuan.

11.2 Academic Evaluation

Selected papers of Wang Yuan was published in 2005. Phua Kok Khoo sent me two book reviews, which were recorded below as evaluation of my whole career:

Wang Yuan was born in the Zhe Jiang province of China in 1930. He graduated from Zhe Jiang University in 1952, and he was assigned by the government to work at the Institute of Mathematics of the Academia Sinica. He joined the number theory section 1 year later, and began to work with Hua Loo-Keng. In the subsequent years, he has published over 70 papers and 11 books. This volume is a selection of 40 of Wang's research papers.

Wang is perhaps best known for his work in sieve methods. The Selberg Λ^2 sieve was a relatively new tool when he began his researches in the early 1950s. Wang was able to combine Selberg's sieve with Buhstab's iteration method to produce a lower bound sieve. Sieve experts will recognize this approach as a precursor to the work of N. C. Ankeny and H. Onishi [Acta Arith 10 (1964/1965), 31–62; MR0167467]. By combining this approach with Kuhn weights, Wang proved in 1957 that if GRH is true, then there are infinitely many primes p such that $p + 2$ is a P_3. (A P_r is a number with at most r prime factors.) After the Bombieri-Vinogradov theorem was proved in 1965, his proof could easily be modified to become unconditional. Of course, the best result in this direction is Chen's Theorem (1973) that $p + 2 = P_2$ infinitely often. Chen's proof uses the linear sieve of Halberstam and Richert, which involves using arbitrarily many Buhstab iterations. In joint work with Ding Xiaqi and Pan Chengdong, Wang showed that his technically simpler sieve (which uses only one or two Buhstab iterations) is sufficient for the proof of Chen's Theorem.

Wang also proved that if F is an irreducible polynomial of degree g, then there are infinitely many integers n such that $F(n) = P_{g+1}$ if $g = 3,4$, or 5. Richert later extended this to all $g \geq 3$, and Iwaniec proved that a quadratic F represents a P_2 infinitely often. Another interesting result of Wang's involves Latin squares. Let $N(s)$ denote the number of mutually orthogonal Latin squares of order s. Wang used sieve methods to prove that if $s \equiv 2 \pmod 4$ and $s \geq 6$, then $N(s) \gg s^{\frac{1}{26}}$. This result was later improved to $N(s) > s^{\frac{1}{17}} - 2$ by R. M. Wilson [Discrete Math. 9 (1974), 181–198; MR0167467].

Other areas represented in this volume include properties of the σ and φ function (including one paper with A. Schinzel), Diophantine approximation (including one paper with W. M. Schmidt), applications of Diophantine approximation to numerical integration (including a series of six papers with Hua), and applications of number theory to applied statistics (including four papers with Fang Kaitai). Overall, this volume forms a nice tribute to one of the significant figures of Chinese number theory.

S. W. Graham (See Math. Rev; 2,169,293, 2007c11002).

The Chinese mathematician Wang Yuan is an influential and important figure in the development of mathematics in modern China. A long-serving assistant and companion to his teacher the distinguished mathematician Hua Loo-Keng, Wang himself was also often at the center of mathematical activities throughout China before the onset of the notorious Cultural Revolution, under which both men suffered worse than intolerable harassment.

The selected papers in the book being reviewed include most of Wang's significant contribution to mathematics. In number theory, there are his papers on the investigation of the Goldbach conjecture, the least primitive root of a prime, Diophantine inequalities for forms in an algebraic number field. The papers in numerical analysis include his joint work with Hua on the use of Diophantine approximations in numerical integration. The value of the book would have been much enhanced had there also been some additional comments on more recent developments at the end of an individual paper, especially since Wang himself is the editor to his own selected papers.

P. Shiu (See Zentr. Math; 1119–15, 2008, 01,026).

Ye Yangbo, professor of Iowa University, went to the Institute for Advanced Study, Princeton in Jan. 2008 to join the memorial program in honor of Selberg and wrote to me:

I had a long talk with Professor P. X. Gallagher. At the sight of your photo, he spoke of your results of (1, 3) and (1, 4) in glowing terms. He told me, since Bombieri and Vinogradov proved the mean value theorem, he had been convinced that you proved (1, 3) and (1, 4). During lessons on analytic number theory, he gave in Columbia University in 1966, he lectured on your unconditional proof of (1, 3) and (1, 4) according to your articles. He praised your proof for clarity, its last steps in particular, compared to other papers on this subject, and hailed it as excellent and beautiful. Moreover, he told me, many of his peers shared the opinion.

I told him that in 1966 you got the point of Professor Chen Jingrun's paper on (1, 2) at the first sight of its abstract and talked about your nobility of character, for which he expressed admiration.

Also he spoke highly of your work on least primitive root of a prime and almost primes represented by integer-valued polynomial.

He said he had the *Introduction to Number Theory* by Mr. Hua and admired the innovative proof of the density theorem, among others, as better than the four proofs previously. I know you've done a great deal for the book but have no idea whether you wrote the chapter or not, therefore I didn't elaborate on this.

I told Professor Gallagher that I would share his kind words with you. He was so pleased and asked me to write and pay his respect to you.

Best, Ye Yangbo.
Jan. 14th, 2008.

Professor Gallagher was a distinguished expert in analytical number theory and tenured professor of Columbia University. The chapter on density in the book he mentioned was written by Hua Loo-Keng and me.

My research was sieve method oriented, therefore, without one mean-value formula when working on (1, a), the (1, 4) and (1, 3) I had were conditional results under the assumption of Generalized Riemann Hypothesis. Pan Chengdong developed and proved some mean-value formula and consequently fully proved (1, 5) and (1, 4). Chen Jingrun took my (1, 3) a step further with his transition theory and finally got (1, 2). For sure they've done a great job.

Pan Chengdong once told me that I made my points clear in my papers, in contrast to some other articles on sieve method. Chen Jingrun told me that (2, 3) was good enough and could not be better. I employed elementary methods in proving (2, 3), however it took more advanced knowledge to prove (1, a).

11.3 Popularizing Science

Research on mathematics had turned too arduous and consuming for my age and physical condition. But I would not call it off. I tried to take up some work to the best of my ability.

Together with Pan Chengdong and Jia Chaohua, I compiled Collection of Hua Loo-Keng's Works (Number Theory), which was published by the Science Press.

With Pan Chengdong I compiled *Selected Papers of Chen Jingrun*, which was published by Jiangxi Education Publishing House.

I translated algebraic number theory classic, *Lectures on the Theory of Algebraic Numbers* by Erich Hecke, which was published by the Science Press in 2005.

With Li Wenlin, I translated Bruce Schechter's book *MY BRAIN IS OPEN: The Mathematical Journeys of Paul Erdos*, which was published by Shanghai Translation Publishing House in 2002.

With Xu Mingwei, I translated *Advanced Calculus* by Loomis and Sternberg, which was published by the Higher Education Press in 2005. I did about 20% of the translation for this one.

In 1995, I paid a 2-month visit to Academia Sinica in Taiwan and spent one week in National Cheng Kung University in Kaohsiung during the visit. Fong Yuen, the host, offered to write a book on calculus with me. I took it for some revisions on his lecture notes and considering this as cooperation between friends, I said yes. Later I read part of his manuscript, which was well-written and in English. The Springer Hong Kong agreed to publish the English version. In this way, I took fruits of his labor for no good reason. In return I took up the remaining work as far as possible, that was to translate the book into Chinese and completed the exercises in the book. Fong Yuen is undoubtedly the one and only author of this book, the English version of which was published by Springer in 1997.

Mu Peng, editor with Shangdong Education Press, has been keen on publishing my popular science writings. I worked with Li Wenlin to collect my works on science and asked him to be my editor. Li classified the articles and the book, titled *Wang Yuan on Goldach Conjecture*, was published in 1999.

Afterward, I wrote some articles to popularize maths and essays of various subjects, which Dalian University of Technology Press planned to sort out and publish. I worked as the editor for the collection, titled at first *Book Notes on Number Theory* and later *Wang Yuan on His Academic Career*.

I took charge of the translation of *Iwanami Dictionary of Mathematics* and *Dictionary of Mathematics* published in the former Soviet Union.

Also I was editor-in-chief for two books published by the Science Press, i.e. *Dictionary of Mathematics* and *Overview of Academic Achievements of Chinese Scientists in the 20th Century*.

I spent a lot of efforts on the *Dictionary of Mathematics*. First of all, I had to organize 17 editing teams, leaders for which would be responsible for working out the entries for their own teams and selecting and confirming writers and referees. The editor-in-chief and associate editor-in-chief were in charge of liaison and coordination with the team leaders, which was conductive to increase motivation. Thanks to the over 300 scientists joined as writers and over 200 as referees, the dictionary had been finished soon and hailed enthusiastically by readers and authorities, the latter offered a fund of 3.6 million yuan for editing of 5 other dictionaries, i.e. mathematics of the second edition, physics, chemistry, earth science and life science. The second edition of *Dictionary of Mathematics* has been released. So far the dictionary has been printed for six times in view of the great demand.

In memory of the centennial anniversary of Hua Loo-Keng's birth, Higher Education Press decided to reprint his *An Introduction to Higher Mathematics*. They asked me to join as the proofreader and I wrote an introduction for the book as a lead-in. The series of four books was published in 2009. Cambridge University Press published the English version.

Zeng Qiwen of Hong Kong University gave me a 50-page lecture note on analytic number theory by Selberg in Hong Kong. I supplemented details and developed it into a 100-page article, which I posted on IAS webpage in memory of Selberg after his death. My maths career started from learning his sieve method.

The latest book I worked on was Part I of *Algorithmic Number Theory*. Feng Keqin and I took charge of translation of the book, which was published by Higher Education Press. Maybe I'll publish some more popular science books later.

11.4 Enthusiast for Calligraphy

When I turned 66 years old, I started to think about picking up hobbies of younger years, which had been long shelved as a result of my devotion to mathematics. Calligraphy seems to be the most accessible because you need a teacher to learn painting. Besides, I joined the violin team of Zhejiang University during college years and at that time felt it difficult because the fingering was different from the Huqin I learned at high school, which, in addition to its high demand for speed and strength in playing, made me feel I was not a good fit for it. In contrast, I could teach myself calligraphy by copying at the beginning and improve gradually. I decided to have a try.

I got started from regular script from September 18, 1996, and worked on Liu's style first and Yan's style later, but rarely on Ouyang's style and Zhao's style. After a while, I appreciated Liu's style more than others and practiced his Xuanmi Tower Inscriptions for many times. But I preferred running script and cursive script and turned to copying from Wang Xizhi's classic *Preface to the Sacred Teachings*, before I turned to the regular script of Dong Qichang and the cursive script of Huaisu and Mao Zedong. It was not until a decade later that I spent half of the time on writing independently and the other on copying.

Like Ouyang Zhongshi, a celebrated calligraphist in China, I was in the CPPCC non-party team from 1998 to 2002. It was fate. In fact, the State Council previously gave a calligraphy scroll to each of the first PhD supervisors in China and the one for me was written by Ouyang Zhongshi. I mentioned the scroll to him, but he didn't remember it. It was fate, as sure as fate. We lived in the same hotel every year during the CPPCC session. I would pay a visit to him at night and take my writings to ask for his opinions. However, it was watching him write benefited me most.

One year we lived in the CCECC Plaza. The hotel manager invited Ouyang to write and he asked me to go together. When he finished his, the hotel manager asked me for another and I completed a horizontal scroll, reading "Great Plans Fulfilled". I didn't write that big characters before that and had no idea of writing with wrist raised, my hands kept shaking when writing, which revealed to me the importance of the way of holding the brush. After I quit my job as CPPCC member in 2003, I had paid visits to Ouyang Zhongshi at Capital Normal University and he had come to CAS for calligraphy activities to offer help for enthusiasts. I watched him write again and his way of holding brush in particular, which helped me a lot. I heard on several occasions Ouyang's brilliant comments on learning calligraphy and I am basically a faithful disciple.

Interest is the best teacher and I never stop practice in calligraphy.

For a professional mathematician, the one and only important task is to study mathematics. I would spend rest of my life reflecting and reviewing my previous researches. As far back as the year 1980, I was invited by *Science*, a Japanese magazine, to write a short biography titled *My Maths Life* and later published *My Memories* and *Analytical Number Theory in China*, all of which were translated into Japanese by Tomiko Shirashima. Those articles were but brief accounts and didn't take in my work after 1980, therefore I wanted to have a more detailed memoir to give an account of my school experience and my career before 2010, when I was 80 years old. I might as well talk about something else. I would take my time to recall, think, and take records.

Printed in the United States
by Baker & Taylor Publisher Services